Pvt. Ltd.

To my beloved father,
mother, and wife

Membrane Separation Process

Advanced Application Based Simulation

Dwaipayan Sen

Department of Chemical Engineering
Heritage Institute of Technology
Kolkata, West Bengal

2018

Studium Press (India) Pvt. Ltd.

Membrane Separation Process

Advanced Application Based Simulation

ISBN: 978-93-85046-16-2

Published by:

Studium Press (India) Pvt. Ltd.
4735/22, 2nd Floor, Prakash Deep Building
(Near Delhi Medical Association),
Ansari Road, Darya Ganj, New Delhi-110 002
Tel.: + 91-11-43240200-15 (15 lines); Fax: 91-11-43240215
E-mail: studiumpress@gmail.com

Printed at India

About the Author

The author is currently being at the Department of Chemical Engineering, Heritage Institute of Technology, Kolkata, West Bengal, India and serving as the Assistant Professor to the department. Prior to this institute the author was at the Department of Chemical Engineering, National Institute of Technology, Agartala, Tripura, India, from where he started his journey as an academician. He has also being in the industry for one and half years before he completed his doctoral degree from the Department of Chemical Engineering, Jadavpur University, Kolkata, West Bengal, India in 2012. In 2003 he had done his bachelor and in 2006, he had completed his masters from the Department of Chemical Engineering, Jadavpur University, Kolkata, West Bengal, India. He is life member in Indian Institute of Chemical Engineers, Kolkata Division since 2012. In his short career he had several publications in peer reviewed journals like RSC Advances, Journal of Membrane Science, Industrial Engineering and Chemistry Research etc. He had co-authored book chapters in publication house like Taylor & Francis and Nova Science. He had jointly authored three contributions to "Encyclopedia of Membranes" (*Editors*: Enrico Drioli, Lidietta Giorno) from a publication house like Springer. Presently his group is working on biofuel, use of membrane in enhancing antibody specificity and membrane application for the separation of polyaromatics from oily wastewater. Currently he is supervising four PhD students. He was recognised through several awards from industry and academics in his short span of career. Recently in 2016 he had received "Certificate of Reviewing" award from Elsevier for reviewing articles submitted in Chemical Engineering Journal.

Acknowledgement

I thank Prof. Chiranjib Bhattacharjee, Professor, Department of Chemical Engineering, Jadavpur University, Kolkata, India and Dr. Kalyan Gayen, Assistant Professor, Department of Chemical Engineering, National Institute of Technology, Agartala, India for giving their valuable suggestions, knowledge and information that helps me to prepare this manuscript. Also I will be thankful to my family, without the support of whom I may not be able to complete this daunting task.

Preface

Although the name of the book comes up with a title - "Membrane separation process: Advanced application based simulation", the book not only tries to derive some dry derivations of mathematics in order to simulate the membrane separation process, but also attempts to ponder upon the philosophy of membrane based separation at different industrial scale. First few chapters in this book will provide you an idea of different membrane separation processes and its wide application at different levels of society. This discussion might help the readers to understand the insight of membrane separation process at different segments of our society. The last few chapters will elaborate the basics of mathematics that may be extended to simulate the membrane separation system. At the end of the book, we had given some codes written in MATLAB programming language.

Dwaipayan Sen, PhD
Assistant Professor
Department of Chemical Engineering
Heritage Institute of Technology
Kolkata

Table of Contents

Chapter 1

Basics of Membrane Separation Process

One of the basic definitions for separation process is separating one or more components from a mixture in their pure form. The most advanced application in such separation technology is the membrane separation process, where an inert barrier is placed across which the separation occurs. Nowadays membrane separation had gained a considerable interest in several industrial applications. Especially, people had shifted from normal microfiltration to nanofibril membrane to have better effective design of the separation process. One of the oldest applications of the membrane separation technique was identified by Abbé Nolet called "osmosis" in 1748, where the permeation of water was observed through diaphragm. However, there were no such industrial applications of membrane until 19th and early 20th century. In 1960, the cellulose acetate reverse osmosis membrane was first used in the desalination plant. In 1980, pervaporation was used for dehydrating alcohol towards its commercialization. In this chapter we will discuss different membrane processes, membrane modules and mathematical concepts related to membrane separation process.

1.1. MEMBRANE PROCESSES' CLASSIFICATION

The classification of membrane separation process might be broadly viewed as porous membrane and impervious membrane. However, again the pervious and impervious membrane can be classified into several processes like microfiltration (MF), ultrafiltration (UF), nanofiltration (NF) and reverse osmosis (RO). Fig. 1.1 describes a schematic representation of different membrane separation process.

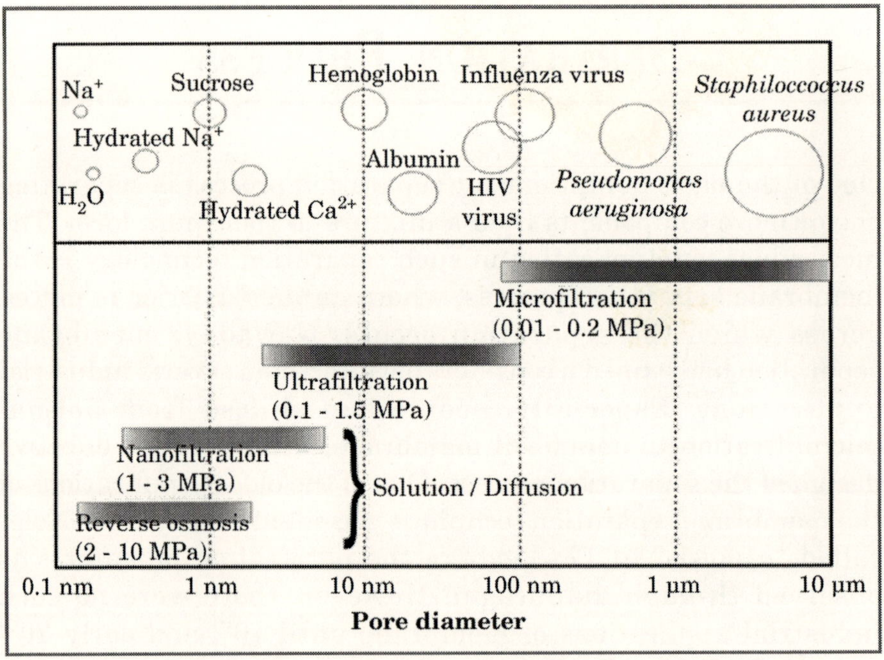

Fig. 1.1: A representation of different membrane separation techniques (*Reproduced from (Tiraferri, 2015) under Creative Commons Attribution 4.0 International License*).

Among these several membrane technologies UF and RO are the most widely used technologies in different applications. In general the UF process is primarily used for the separation of higher molecular weights like whey proteins, antibodies, lignosulfonates etc. While RO is mainly limited to desalination plant, where ions are rejected using membrane and fresh water is accumulated. MF and UF are porous membranes, while on the contrary NF and RO are non-porous membranes. However, NF is at the intermediate level, where the NF near to UF range are porous membrane and NF close to RO sides are nonporous membrane. Again, based on the fabrication of the membrane there are two types – one is called symmetric membrane and the other one is called asymmetric membrane. Fig. 1.2 shows the images of two types of membrane.

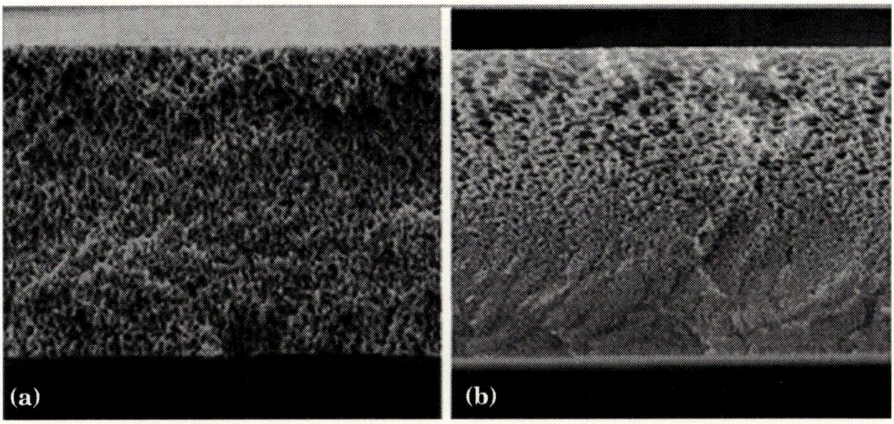

Fig. 1.2: (a) Cross-sectional image of symmetric membrane (b) Cross-sectional image of asymmetric membrane[1]

Now, if you look at the image, symmetric membrane is having uniform porous structure across the thickness of the membrane. It is also called isotropic membrane. While, on the contrary asymmetric membrane does not have the uniform porous structure across the thickness and called anisotropic membrane. Therefore, as seen in Fig. 1.2 (b), there is some non-uniformity across the thickness of the membrane showing the asymmetric nature of the membrane. In our subsequent sections, we will discuss on different membrane separation processes, characterisation of the membrane,

[1] Source: http://www.cemag.us/article/2007/07/optimized-particle-filter-copper-plating-solution

what are possibilities of membrane materials and the most important one, what should we look after, when we are planning to buy a membrane.

1.1.1. Microfiltration (MF)

Microfiltration (MF) is a process for removing particles or biological entities in the range of 0.025 µm to 10.0 µm from fluids after passing through a microporous membrane. Micron-sized particles can also be removed through non-membrane filter like fibrous media such as clothe, but only membrane filter can precisely separate the materials based on defined pore size and thus can ensure the quality. Membrane filters can be used for either as a tertiary treatment process or might be a prefiltration prior to other filtration having less pore diameter compared to MF. However, sometimes MF and cloth filtration can complement each other, especially when the feed to the membrane is much suspended solid borne. Along with the use of MF membrane as the clarifier, fluids containing bacteria can be filtered using MF in order to arrest microorganisms on the membrane surface for subsequent culture and analysis. Another crucial application of microfiltration is during the preparation of sample obtained after cell rupture to separate the cell debris from the lysate. Membrane pore size ranges from 0.05 µm to 1.0 µm. MF membranes are fabricated using various materials like, polysulfone, polyvinyldifluoride (PVDF), polyethersulfone (PES), ZrO_2 and carbon. They are available in the market with different configurations like pipe shaped membranes (hollow fibre or tubular) and plate shaped membranes (plate type). The details of different membrane module configurations are elaborated later.

1.1.2. Ultrafiltration (UF)

Ultrafiltration (UF) is a process of separating smaller particles, generally in the form of miscible one into the solvent. The basis for the separation of the molecules is their molecular size difference, the permeability of the membrane, which might get affected with chemical and physical interactions between the molecules and the membrane materials. In general the specification unit for the molecular weight of the molecule is given as Dalton (Da) and it is understood that the more molecular weight attributes to more molecular diameter of the molecules. Based on this concept, molecular weight cut-off (MWCO) of the membrane is defined as the molecular weight of the molecules, 90% of which is retained by a membrane is called the MWCO of the membrane. For example if we are saying that the MWCO of a membrane is 50,000 Da, then we can say that 90% of the molecules having molecular weight 50,000 Da will be retained by the membrane. At the same time, we can also say that the molecules less than this value of the molecular weight will pass through the membrane and the higher than that of 50,000 Da will be retained by the membrane. However, the case might be different in some cases. Say we want to separate a protein A of molecular weight 66 kDa from other protein B of molecular weight 18 kDa. In that case we may opt 50 kDa membrane and it is expected we should not have any trace presence of protein A in the permeate. While, the expectation might be different in case of actual. We could have some trace concentration of protein A. This is because of the uncoiling of protein leading to a conversion of tertiary structure to primary linear structure. Because of this, in a pressure driven process like UF, the linear structure may pass through the pores towards the permeate side. Fig. 1.3 gives a view on this.

UF generally separates those molecules, among which we have at least an order of magnitude in size. Materials having size in between 1 kDa to 1000 kDa molecular weights are retained, while other low molecular weights will pass through. For example colloidal and particulate matter might be retained on UF membrane. UF is typically used to separate proteins from their mixture, desalting, or its concentration. For example to prepare whey protein concentrate, UF process with different MWCOs is a state of art process, where different proteins, carbohydrates, salts etc. are sequentially separated out. In general protein separation is carried out with 3 kDa to 100 kDa MWCO membrane. There are several advantages incurred with the UF process, which is given below:

- UF is much gentler and non-reactive to solutes compared to the other processes like clarifier, floatation etc.
- UF is considered to be more efficient as it can simultaneously concentrate and desalt solutes with high precision of separation.
- It does not require a phase change, which often denatures species like protein, and can be performed at any temperature, unless the membrane materials itself gain thermal stress.

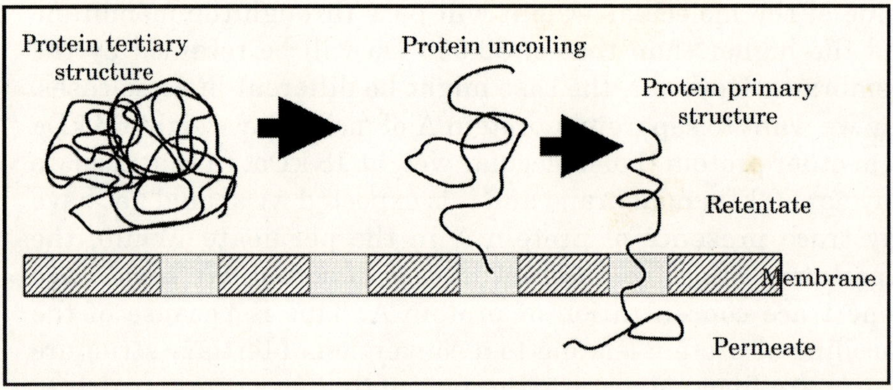

Fig. 1.3: Permeation of high molecular weight protein because of uncoiling or denaturation

1.1.3. Nanofiltration (NF)

Nanofiltration (NF) is a pressure driven membrane process, also known as "membrane softening", has been successfully used for treatment of hard, high colour, and high organic content water. It is also a pressure driven process and hydraulic pressure is used to cross the barrier of feed solution's osmotic pressure. Once the barrier will be overcome, the low molecular weight diffuses through a semi-permeable NF membrane and this process of diffusion follows Fick's law. In general NF membrane has lower monovalent ion rejection properties, making it more suitable to treat waters with low salinity. NF membrane also works as an absolute barrier for cysts and most viruses. Because of its monovalent ion rejection property, NF membrane is specified in terms of percentage sodium chloride (NaCl) rejection instead of MWCO. However, NF shows high rejection for multivalent ions also such as *e.g.*, calcium, magnesium, sulfate, iron, arsenic, etc., and is sometimes referred to as membrane softening. NF is mostly employed to remove the hardness of brackish groundwater in order to produce potable water. NF plants typically operate at 85 to 95% recovery. In subsequent reverse osmosis (RO) stage brackish water RO plants typically show 70 to 85% of the source water into permeate, and seawater RO recovery rates range from 40 to 60%. The *recovery* says about the percentage of the feed we get as permeate and this is exactly what we are doing with the NF process. As said, we are removing the hardness of water prior to RO so that we can get soft water as permeate. Therefore recovery percentage of water as permeate will be a crucial parameter that dictates the feed for RO plant. Water with high concentrations of calcium or magnesium and monovalent salts can be treated with NF prior to a RO, so that the RO will face some low load during its operation. Now if we will look, why we should use NF? The answer must be its low installation land. NF requires an equivalent footprint compared to brackish water RO, and a minimal operational installation space compared to thermal desalination technologies.

1.1.4. Reverse Osmosis (RO)

Reverse osmosis (RO) is process, where pretreated source water rich with ions is fed to the membrane module and the separation happens at a moderate pressure against a semipermeable membrane. The membrane rejects most of the ions and minute sized molecules, and allows water of very low mineral content through the membrane. RO systems have been successfully used to purify saline ground water, brackish waters, and seawater.

Further they are also used for the removal of inorganic contaminants such as radionuclides, nitrates, arsenic, and other contaminants such as pesticides from run-off water. Now, why we are calling it “Reverse Osmosis”? Let us first start with osmosis. What is osmosis? Say we are having some fresh water stored in balloon like enclosure. The only difference is that the material of the balloon is semipermeable membrane. Now we have a brine solution, where we are keeping this balloon filled with fresh water. What we can see? The balloon will shrink as the water in the balloon goes to the solution in order to lower the concentration of the solution. RO is just exactly opposite to this. We are applying pressure and forcing the solvent to move from higher concentration to lower concentration against the osmotic pressure difference. Fig. 1.4 shows a picture on what RO means actually. The equation to this is given as

$$J = \frac{\Delta P - \Delta \pi}{\mu R_{tot}} \tag{1.1}$$

Where, J is the solvent flux through the membrane (L/m^2s), “P is the pressure difference across the membrane (kgf/cm^2), “p is the osmotic pressure difference (Pa), μ is the viscosity of the solution (Pa.s) and R_{tot} is the total resistance (resistance due to membrane plus resistance due to fouling of the membrane) over the membrane (m^{-1}). The osmotic pressure is a polynomial function of the concentration of the solution and is given as,

$$\pi = a_0 + a_1 c + a_2 c^2 \tag{1.2}$$

The constants ‘a_i’ (i=0,1 and 2) is depending on the solute molecule and c is the concentration of the solute either at permeate side or at the feed side. Now as the permeate side of the process is low concentration side and the retentate side is the high concentration side, hence the osmotic pressure will act from the permeate side to the retentate side. Therefore, we can write,

$$\begin{aligned} \Delta\pi = \pi_p - \pi_r &= \left(a_0 + a_1 c_p + a_2 c_p^2\right) - \left(a_0 + a_1 c_r + a_2 c_r^2\right) \\ &= \left(c_p - c_r\right)\left[a_1 + a_2\left(c_p + c_r\right)\right] \end{aligned} \tag{1.3}$$

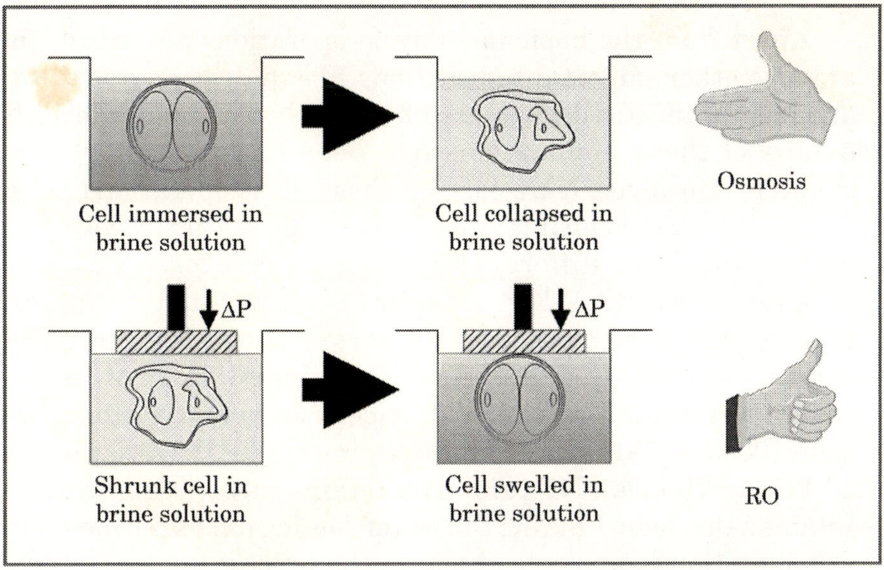

Fig. 1.4: Schematic representation of osmosis and RO with a cell immersed in brine solution

There is another parameter based on which NF and RO membranes are specified and this is called the percentage of salt rejection. The rejection percentage is defined as the amount of solutes being retained by the membrane compared to the feed concentration and is given as,

$$R = \left(1 - \frac{c_p}{c_f}\right) \times 100 \tag{1.4}$$

The primary energy requirement for RO is to apply high pressure at the feed side to pressurize the system, to facilitate permeation. Now depending on the characteristics of the water to be treated, different types of membranes are used. For example in case of seawater desalination (approximately with 3.5 wt% of salt), requires membranes with salt rejections of more than 99.3% to produce an acceptable permeate. Cellulose triacetate membrane membranes are generally used for some seawater desalination plants as salt rejections of about 99.5% with a seawater feed can be attained with this and will have reasonable fluxes.

Apart from the basic membrane operations described, there are also other important operations like dialysis, pervaporation and membrane distillation used frequently in the plant level. The details of these applications had been discussed in the later chapters. However, if we can see these basic operations, we will find all these are in an integrated way applied to purify wastewater and should have different by-products recovered out of this integrated operation. Fig. 1.5 shows an integrated operation for treating casein whey coming out of confectioneries. The effluent coming out of this industry is enriched in proteins and carbohydrates. MF retains high molecular weight proteins (more than 100 kDa), fats etc., UF retains proteins (in the range between 10 kDa to 70 kDa), NF primarily retains carbohydrates and RO retains salts, ions. Finally, the water coming out as permeate from RO operation is again fed back to the process and called ‘green water’ free of any contaminants.

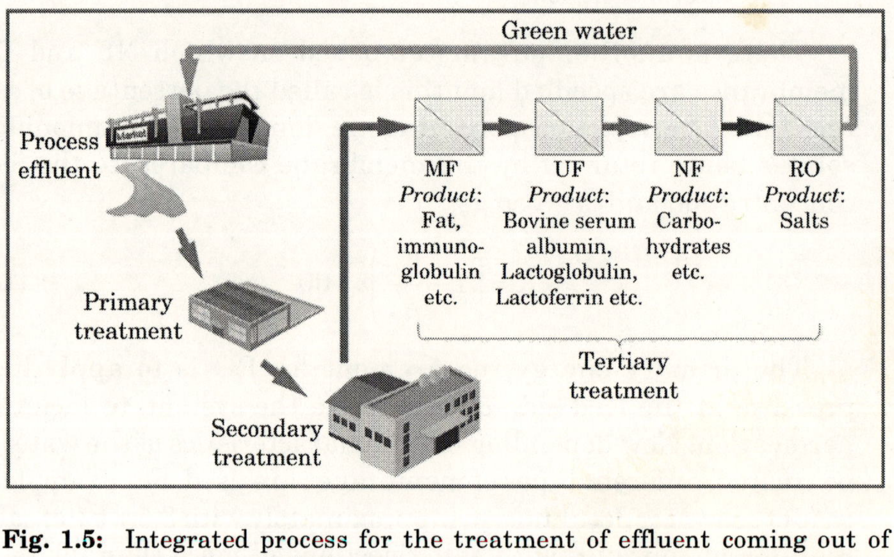

Fig. 1.5: Integrated process for the treatment of effluent coming out of confectionary industry

1.2. MEMBRANE PROCESS AND ITS CHARACTERISATION

As said before, membrane separation is a process, which is

primarily depending on the applied pressure difference across the membrane. Therefore, the membrane separation is a pressure driven process. However, with the process, there is a deposition of the solutes over the membrane surface, which creates a secondary layer over the membrane surface. This layer acts as an impervious layer and increases the deposition possibility with the process. With the onset of the process, the solute concentration over the membrane surface gets increased. This increases the diffusional resistance against the solutes' advance into the membrane and there is sudden decrease in the permeate flux from the membrane. This is called 'concentration polarisation' and falls under the category of reversible fouling. Once the membrane gets cleaned using the specified solution, this layer will be washed off and hence, we can revert back to the original state of the membrane. With the process, as said, this deposition increases and forms a secondary layer called 'gel layer'. This gel layer is the ultimate stage of the concentration polarisation and reduces the flux drastically. However, with the formation of the gel layer, if the process begins to be continued, due to applied pressure, the solutes can go into the pores of the membrane. This is called the clogging of the pores. Once, the pore gets clogged, it becomes difficult to regain the original state of the membrane by simple wash. In that case, we need back-wash of the membrane. That is, with the upside down the cleaning solution will be charged to the membrane and forces the solutes blocked the passage to come out of the pores. However, this process of back-washing is limited to symmetric membrane, where both the sides of the membrane having uniform texture. Therefore, this is called irreversible fouling of the membrane. Fig. 1.6 shows a schematic of both the fouling.

In case of membrane process, the governing equation (eq. 1.1) has a resistance term, which comprises of the summation of all the resistance terms including the resistance term imparted by the membrane material itself. As the contribution of the resistances due to fouling gets increased, increase in the pressure difference will not affect the steady state permeate flux. Now before saying

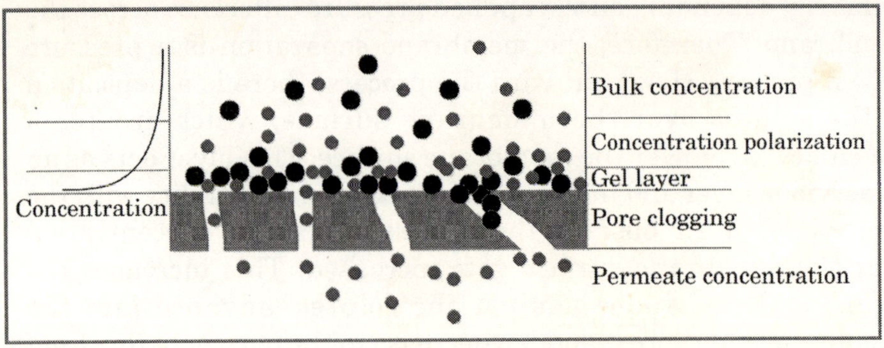

Fig. 1.6: Schematic representation of different types of fouling and concentration profile

anything, let us first elaborate what is steady state permeate flux. Once, the process is having the concentration polarisation, the solute molecule gets diffused away from the membrane surface. In this way, the movement of the solutes towards the membrane surface gets restricted and the flux becomes steady, *i.e.*, there is no further decrease in the permeate flux. As the pressure difference across the membrane gets increased, the value for steady state flux is also getting increased. Further increase in the pressure increases the accumulation over the surface and there is no control of the pressure on the flux. This is when the pressure driven process is shifted to mass driven process. Fig. 1.7 shows the steady state flux versus the pressure difference across the membrane.

The membrane resistance is calculated with the virgin membrane after filtering deionised water at different pressure. This will yield the value for the resistance using equation 1.1, called the resistance for membrane material, which is also called 'membrane hydraulic resistance'. As we don't have any solution to be separated with the membrane, therefore the osmotic pressure difference becomes zero here and the equation yield a linear relation between permeate flux with pressure difference across the membrane. The slope provides us the value for inverse of the membrane resistance, where the inverse can also be expressed as the permeability per unit thickness of the membrane time of the solution of the viscosity and given as,

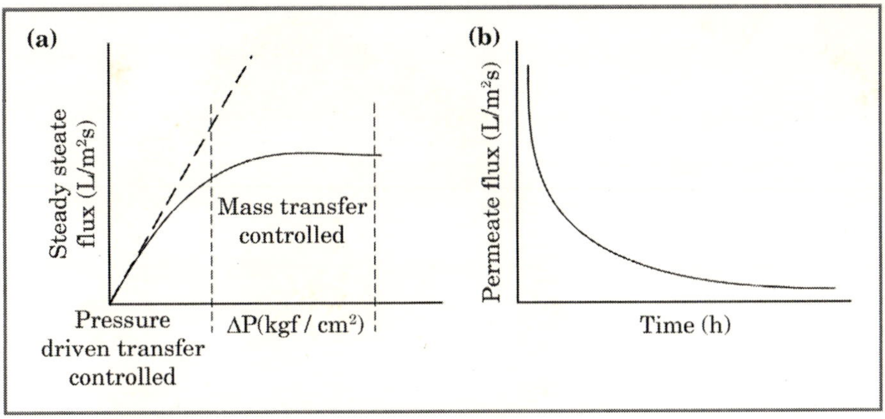

Fig. 1.7: (a) Steady state flux against pressure difference across the membrane (b) Permeate flux versus time

$$J = \left(\frac{P}{l}\right)\Delta P = \left(\frac{1}{\mu R}\right)\Delta P \tag{1.5}$$

Where, P is the permeability of the membrane given as the product of the diffusivity of the solute through solvent (D) times the solubility of the solute (S). Now, the procedure of calculating membrane hydraulic resistance is also called the compaction of the membrane. Now, what do we understand by this compaction technique?

The polymeric membrane is stretchable as the polymeric membrane shows elasticity. During the fabrication of the membrane, we are following the say through phase inversion technique. The porosity of the membrane is controlled through the preparation of polymer concentration. However, we have some distribution of the pore sizes across the membrane surface. Say we want to prepare 0.2 μm membrane, the size of some of the pores might be less than 0.2 μm and some of the pores might be higher than that of 0.2 μm. Therefore, we will have a wide pore size distribution. We are saying the membrane quality is good based on this distribution graph. More and more wideness in the distribution means the quality is bad, while less distribution attributes to good membrane. The compaction, which is done at the highest tolerable pressure of the membrane material, ensures the reduction in this wideness and all the pores to be at the average specified diameter. Fig. 1.8 shows the distribution and its condition after compaction process.

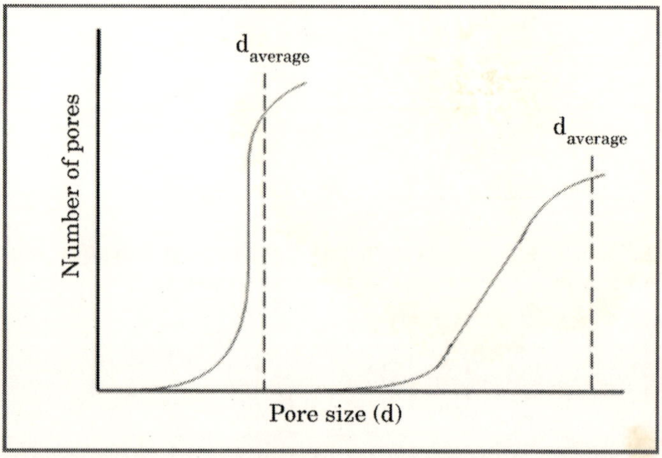

Fig 1.8: Pore size distribution with number of pores

1.3. MEMBRANE FABRICATION

One of the important issues with the membrane separation process is the fabrication of a membrane. The precision of chemical components used during the fabrication, the mechanical procedure adopted are controlling the actual texture of the membrane and the distribution of the pores across the membrane surface area as discussed in the previous section. In this chapter, we will stick to a very basic preparation of polymeric and inorganic membrane.

1.3.1. Polymeric Membrane Fabrication Using Phase Inversion Technique

The primary key controlling parameter for the fabrication of polymeric membrane is the ratios of solvent and non-solvents. A bit tricky in the nomenclature, isn't it? In this case, solvent means, where the polymers can be dissolved, while non-solvent refers to a liquid where polymers are immiscible. Now, what is the basic principle behind the use of such mixture and what should be the properties of the components based on which we can choose? During the preparation of a polymeric membrane, one of the typical procedures is called phase separation. The polymer solution is called the casting solution. Now if this casting solution is exposed to an atmosphere, which is having both non-solvent and solvent. The atmosphere is saturated with the solvent and hence, there should not be any transfer of the solvent either from solution to atmosphere or vice versa. While the non-solvent diffuses through the solution and the polymer is precipitated on the casting film. This process is used primarily for the fabrication of symmetric membrane and called "*precipitation from vapour phase*", where both the sides of the membrane is having equivalent pore diameter. However, the process is difficult to control. Rather, the process might be altered a bit. In the modified one the casting solution is exposed to a coagulation bath, which is prepared with non-solvent. Once, the casting solution over the support is moved through the bath, the non-solvent diffuses from the solution to the bath atmosphere. On the contrary, the non-solvent diffuses through the casting solution and reduces the polymer solubility as because lowering concentration of the solvent in the solution. The polymer gets precipitated and the process is called "*immersion precipitation*". Fig. 1.9 shows a diagrammatic representation of the process.

In the first bath, a concentrated layer of polymer will be formed at the interface between non-solvent and polymer. A thin surface layer is formed. The second bath manifests the actual coagulation and formation of the final film of the membrane. Further, type of solvent in polymer solution is also responsible for the membrane structure and performance as well. For instance, if we are adding non-solvent along with the volatile solvent to the casting solution, it will change the liquid–liquid phase separation process and consequently, will also change membrane the permeability and

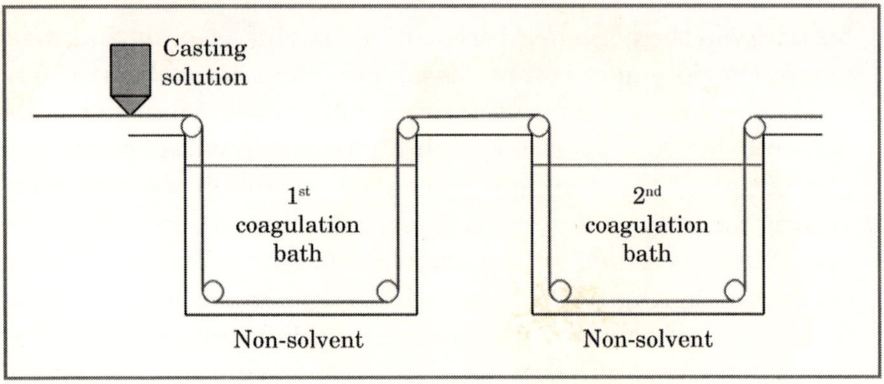

Fig. 1.9: Immersion precipitation method

selectivity. Moreover, using a mixture of solvents in the casting solution will remove the large number of pore formation during instantaneous phase separation, like in case "*precipitation of vapour*". The membrane fabricated in this way called "*phase inversion membrane*". One of the basic properties of the solvent and non-solvent should be that the polymer should be completely dissolved in the solvent and at the same time the vapour pressure of the solvent should be low enough in order to restrict the rigorous evaporation of the solvent so that formation of the large pores (as said before) could be avoided. At the same time the diffusivity of the non-solvent through the selected solvent and the casting solution should be high enough in order to have proper diffusion. Table 1.1 says about the vapour pressure information of some

Table 1.1: Properties of some of the solvents and non-solvents used for membrane preparation

Solvents/ Non-solvents	***Vapor pressure (kPa) at 20°C***
N-Methyl-2- pyrrolidone (NMP)	0
Tetrahydrofuran	19.3
Dimethylformamide	0.3
Dimethylsulfoxide	0.1
Morpholine	0.9
Butanol	0.8
Water	0.62

selected solvents and non-solvents frequently used in the fabrication of membrane. In this chapter, we have confined ourselves to the fabrication of polymeric membrane and tried to give you a very basic of the fabrication. There are other membranes, like ceramic membranes for which the fabrication techniques are bit different and discussed below.

1.3.2. Fabrication of Inorganic Membrane and Characterization

One of the critical issues in maintaining the pore size of the inorganic membrane is the average size distribution of the particles we are using in the membrane fabrication. In general, the fabrication process for inorganic membrane comprises of two sequential steps – one, the fabrication of support layer and the fabrication of top layer. The first one is the mesoporous layer and the later one is the microporous. In this section, we will look after the fabrication of alumina (Al_2O_3) membrane. For the preparation of the membrane support it is seen that if we are using 0.5 µm of Al_2O_3 powder it will yield 0.1 to 0.2 µm of pore sizes. With this, 30 to 50% of the membrane porosity could be achieved.

In the preparation of Al_2O_3 support layer the powder is pressed into a rigid die after applying pressures within a range of 300–400 bar along the single axial direction (Vercauteren *et al.*, 1998). In the subsequent step, this pressed sample is further pressed to a 650 bar and this process is called shaping. After shaping, the samples are subsequently calcined at 500–600°C to release all the organic binder materials and gaseous products that are formed during decomposition and oxidation. In the next step, which is called the presintering that occurs at a temperature between 1100 to and 1200°C, where the neck growth between the Al_2O_3 occurs to provide the products mechanical strength. To fabricate the top membrane layer, a thin film of zeolite is casted on the support. During the fabrication the growth of zeolite is ensured through a hydrothermal treatment, where initially the support id dipped into a standard solution of sodium hydroxide, tetrapropyl ammonium hydroxide and silica in water. After that the growth is autoclaved between a temperature of 80 and 200°C. After film growth, the membranes are calcined at 400–600°C to remove the organic templates.

Once the membrane is being formed the primary issues with the membrane is the characterisation of the membrane. Three basic parameters that are more rigorously discussed are – (a) pore size distribution (b) molecular weight cut-off or the rejection of the membrane and (c) hydrophilicity and hydrophobicity of the membrane. The pore size distribution is a characteristic by which we can say how good the membrane is? This is something a Gaussian distribution around the average diameter of the pores of the membrane. If we say that we want to fabricate a membrane of pore diameter 0.2 μm, we must have a sharp distribution around the value. If we have wider standard deviation or span of the curve we will conclude that the membrane is bad. However, this characterisation is limited to the porous membrane. Fig. 1.10 shows the pore size distribution during the fabrication of membrane, which helps to determine on the goodness of the membrane.

Now, if you look at the figure, we want to fabricate the membrane of pore diameter around the dotted line. In this respect, we can say A is the best membrane as we don't have too wider distance from the average pore diameter in the distribution. While B is also having average pore dimension close to the desired, but the distribution is wider. Wider means, we may have significant number of pores for which dimension is significantly larger than the average one and also significant number for the smaller ones compared to average diameter. C and D is poor, as we don't even reach to the average dimension of the pores and maximum numbers of pores are of fewer dimensions.

For MF and UF membranes, generally we will speak about the molecular weight cut-off (MWCO) of the membrane, which is defined as the molecular weight of the solute, whose 90% is rejected

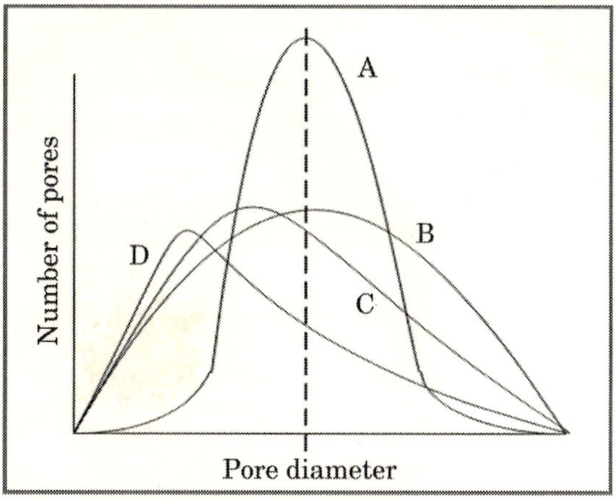

Fig. 1.10: Pore size distribution of a membrane

by the membrane. Let me explain this. Say we have a mixture of A and B dissolved in a solvent C. The molecular weight of the A is 100 kDa[2], for B it is 20 kDa and for C it is 0.1 kDa. Now if we will see that 90% of A is retained by the membrane, then we can say the MWCO is 100 kDa. The below figure gives an idea of finding out the MWCO of the membrane. Fig. 1.11 shows the calculation of MWCO of the membrane.

The rejection of the membrane is said as,

$$R_{obs} = 1 - \frac{C_{sol,P}}{C_{sol,F}} \tag{1.6}$$

Where, $C_{sol,P}$ is the permeate concentration of the solute, $C_{sol,F}$ is the bulk concentration of the solute and R_{obs} is the observed rejection coefficient. When, we have to find out the value for real rejection coefficient and is given as,

$$R_{True} = 1 - \frac{C_{sol,P}}{C_{sol,M}} \tag{1.7}$$

[2] Dalton (Da) is the unit for MWCO of the membrane, which is equal to g/ mole.

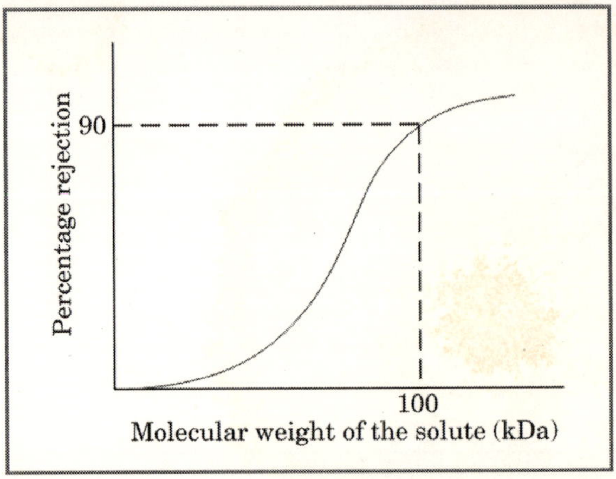

Fig. 1.11: Calculation of MWCO of the membrane

Where, $C_{sol,M}$ is the solute concentration over the membrane and difficult to measure. Because of this, the rejection is called true rejection. This method for the determination of the MWCO is called as "*solute rejection method*". The relation between the real and the observed rejection is given as,

$$\frac{R_{obs}}{1-R_{obs}} = \frac{R_{True}}{1-R_{True}} \exp\left(-\frac{J}{k}\right) \tag{1.8}$$

Where, k is the mass transfer coefficient for which the value can be evaluated from a Dittus Boelter type equation correlating Sherwood number with the Reynolds and Schmidt number as given below.

$$Sh = a\,Re^{b}\,Sc^{c} \tag{1.9}$$

a, b and c are the regression constants and the values of these constants depend on the module and the process. In case with the RO and NF membrane the specification is given by salt rejection. In case with the RO membrane, the membrane is completely rejecting salt. However, for NF, this is not true. NF membrane is not a complete barrier to the salt, rather depending on the salt nature and the condition of the solution it acts as a barrier to the salt. Now, all these I had discussed so far stick to evaluate the membrane efficiency. While, to understand the process efficiency the most common terminology is "*recovery*", which is defined as the percentage of permeate collected out of the feed supplied to the membrane system. Now, on what basis we should select the membrane so that the recovery could be higher. Recovery lower attributes to the two facts or we may say it depends on two conditions – one, the nature of the membrane and the second one is the cake deposited over the membrane. The first one is something we are choosing after evaluating the feed that we want to introduce to the process. Say we are using oily water, where oil needs to be separated out. If we want to reduce the deposition or sticking of oil onto the membrane surface that means we require a membrane surface that attracts water and repels oil. That means the surface must be water loving or "*hydrophilic*". For example poly-vinyl-difluoride (PVDF) membrane is hydrophilic. The reverse is called "*hydrophobic*", where water will be repelled. Like poly-ether sulfone (PES) membrane is the hydrophobic one. This nature of the surface is evaluated based on the water contact angle measurement. If the contact angle is less than 90° the surface is called hydrophilic and if it is more than that of 90°, it is called hydrophobic. Fig. 1.12 shows two surfaces' contact angles.

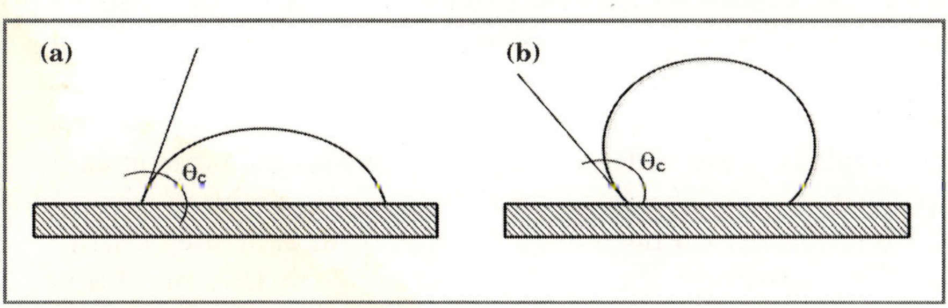

Fig. 1.12: (a) Hydrophilic membrane (b) Hydrophobic membrane

With a batch or semi batch membrane separation process, where we want to concentrate something, the concentration on the feed side increased. This increases the possibility of concentration polarisation and the process gets hindered. Also the recover becomes lower. This volume reduction in the feed side is measured after calculating the ratio between final volume and initial volume on the feed side. This ration is called *"volume concentration factor (VCF)"* (equation 1.10).

$$\text{VCF} = \frac{\text{Final volume on the feed side}}{\text{Initial volume on the feed side}} \tag{1.10}$$

If the VCF is too high, the deposition occurs spontaneously and the recovery drops. In that case to avoid such scenarios or to increase the purity of the concentrate, we are making up the reduced feed volume with the solvent and continue the process. This increases the recovery as well as the purity of the concentrate on the feed side. The technique is called *"diafiltration"* and a very common in the MF/UF range. Now, another important aspect of the membrane separation process is the membrane module, where the membranes are fitted and process efficiency depends a lot on the selection of the module, which are discussed in the next section.

1.4. MEMBRANE MODULES

A membrane module comprises of a membrane casing holding the membrane, across which the pressure is applied in order to separate the solutes from the solvent. In general, the modules can be differentiated based on the modus operandi of the process. One is called the dead-end module and the other one is called cross-flow module. In case of dead-end membrane module the pressure is applied to the membrane normally and correspondingly, the feed flow direction is also normal to the membrane. While, with the cross-flow operation the flow is tangential to the membrane. Therefore, the possibility of deposition over the membrane surface, which ultimately creates a secondary impervious layer over the membrane surface, is less in cross-flow. On the contrary, with the dead-end module, as the direction of the flow and applied pressure coincides, there is a high probability of membrane fouling due to pore clogging. Hence, in industrial operation, cross-flow is far superior compared to dead-end operation. Fig. 1.13 shows the schematic representation for these two types of operations.

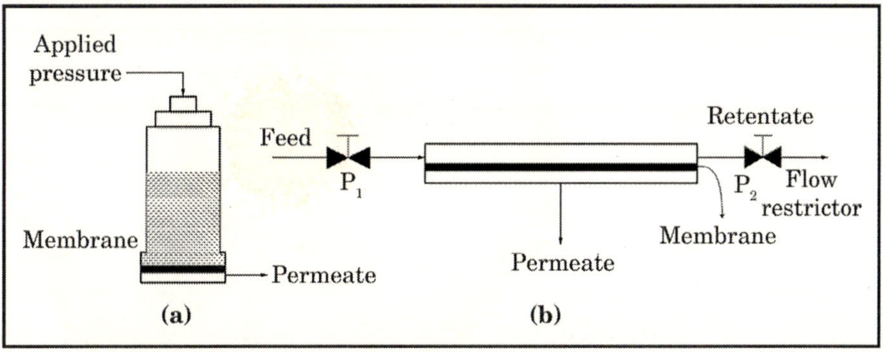

Fig. 1.13: (a) Dead-end membrane module (b) Cross-flow membrane module

In case of dead-end module the pressure difference across the membrane is given as $\Delta P = P_{appl} - P_p$, where P_{appl} is the applied pressure for this batch process. P_p is the pressure at the permeate side and usually taken as the atmospheric pressure. Therefore, the gauge pressure reading will be considered as the applied pressure difference across the membrane.

With the cross-flow system as shown in the above figure, pressure P_1 is the pressure head developed due to the delivered pressure by the pump. In other word we may call this as the dynamic head. P_2 is the back pressure created with a flow restrictor fitted at the outlet of the membrane module. Therefore, the pressure difference across the membrane can be said as $\Delta P = P_{average} - P_P$, where $P_{average}$ is the arithmetic mean of P_1 and P_2. Considering the same value for P_p as in the previous case, the average of pressure gauge reading for the inlet and outlet of the membrane module will be taken as the pressure difference across the membrane. There are different types of cross-flow membrane modules, some of which are discussed elaborately in the subsequent sections.

1.4.1. Plate-and-Frame Membrane Module

In this module the membrane plates are stacked separated by a frame like structure. If you see at the Fig. 1.14, the arrows show the upstream and permeate paths. The upstream leaves the module as the retentate. Permeate is collected from channels in support plates. One of the typical advantages is its easy cleaning and replacement of membranes. However, low membrane area per volume of the module (packing density) and this yields low recovery. At the same time, the module needs several mechanical sealing, which might fail in extreme case. In general the module has the packing density of around 100-400 m^2/m^3. The plate-and-frame module is much efficient with pervaporation set-up. However, because of its low packing density, it is much less popular compared to spiral wound and hollow fiber modules, which are used for gas separation applications.

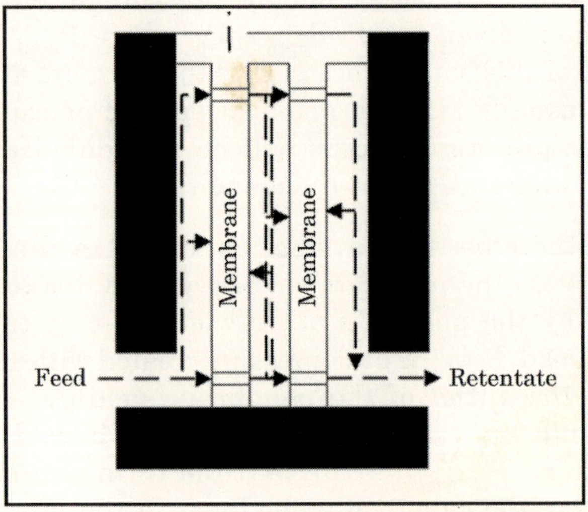

Fig. 1.14: Plate-and-frame membrane module

1.4.2. Spiral Wound Membrane Module

This wound module is derived from a flat membrane or we may

say that it is a modified version of a flat sheet membrane (Fig. 1.15). Here, the membrane is wrapped around a perforated collection tube and between the membranes there is a spacer, who not only maintains the separation of two membranes to be stuck with each other, but also it acts as the turbulence promoter to reduce fouling. It may be said that it is a modified version of a plate-and-frame system wrapping a central collection pipe, like a sandwich roll. The figure shows that the feed flows axially over the envelopes (a cassette formed by stacking the membranes between two spacers – one at the feed side and the other at the permeate side built up by two membrane stacks. The permeate gets collected into the central permeate collection pipe after being moved through envelop. Packing density of this module (300–1, 000 m^2/m^3) is more than that of the plate-and-frame module. However, the performance of the module strongly depends on the channel height and it is determined by the thickness of the permeate and feed-side spacer deposition. The module is easy to operate and it is inexpensive when the question comes on changing the feed spacer thickness to overcome concentration polarization and fouling. However, still the low packing density is a major issue in membrane separation.

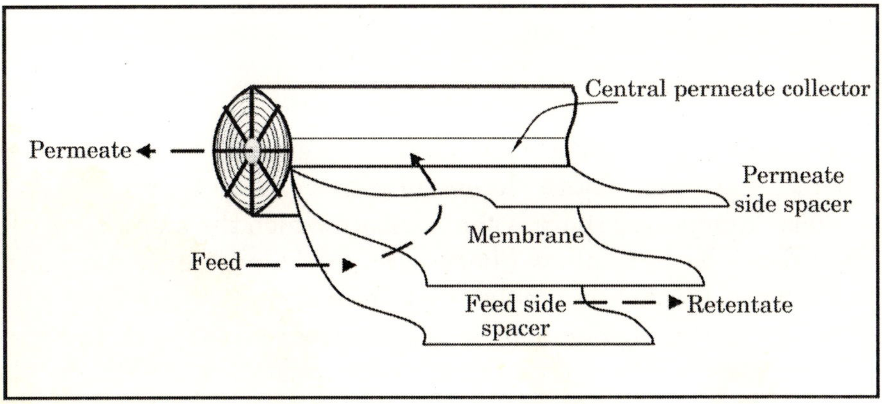

Fig. 1.15: Spiral wound module

1.4.3. Tubular Membrane Module

Tubular membrane module (Fig. 1.16) depends on the availability of the tubular shaped membrane. In general the tubular membranes in the range of 4 to 18 numbers are being clubbed and put in a steel casing of diameter more than 10 mm. The packing density is quiet a low for the module, around 300 m^2/m^3. The feed solution goes into a central tube attached to the membrane. The permeate comes out through the membrane and is collected. In general, the membrane is much to the ceramic membrane.

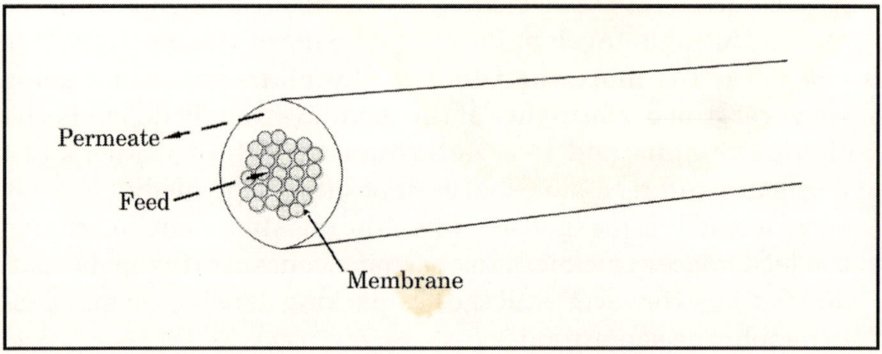

Fig. 1.16: Tubular membrane

1.4.4. Hollow Fiber Membrane Module

The hollow fiber membrane module is similar to shell and tube heat exchanger, where the membranes can be thought of as the tubes (lumen) and there is shell, within which the membrane tubes are fixed. Either hollow fibers (<0.5 mm diameter) or capillaries (0.5-5 mm diameter) or tubes (5-15 mm diameter) are fixed within the bore or shell. The ends of the fibers are fixed with the casing using agents like epoxy resins, polyurethanes, silicone rubber, thermoplastics, thermosets, or inorganic cements. In some of the cases, they are fused by heating. The packing density for hollow fiber membrane module is very high, almost 30,000 m^2/m^3. This is one of the primal advantages with this module and also it has the lowest manufacturing cost. However, with the module there might be excessive pressure drop. To avoid such excessive pressure drop, large diameter fibers may be required in some cases. It is seen that more than 80% of the membranes used for gas separation are hollow fiber membrane module. There are two types of module arrangements – one, where the feed flows through the lumen and the permeate is collected at the shell. This is called "*inside feed*" (Fig. 1.17(a)). In the second arrangement, the feed enters at the shell side of the hollow fibers, while the permeate flows through the lumen. This arrangement is called "*outside feed*" (Fig. 1.17(b)).

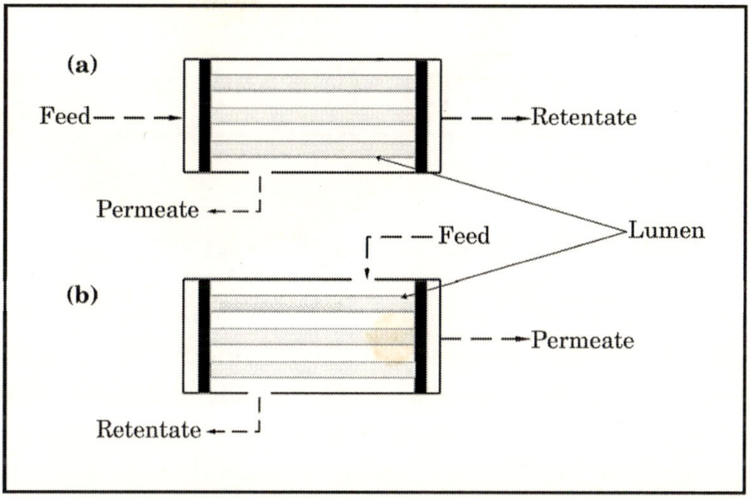

Fig. 1.17: (a) "Inside feed" hollow fiber membrane (b) "Outside feed" hollow fiber membrane

1.5. CHAPTER IN BRIEF

- Basics of membrane separation process.
- An elaboration of different type of membrane separation process.
- Different modes of membrane separation process.
- Different parameters for membrane characterisation.
- Different types of membrane modules.

Chapter 2

Application of Membrane in Wastewater Treatment

2.1. OIL BASED INDUSTRIAL WASTEWATER

Wastewater treatment in petroleum refineries is one of the intricate processes, which challenges environmental management because of versatile by-products – both volatile and toxic. Hence, petrochemical wastewater requires a combination of treatment methods to remove oil and other contaminants that are varying in nature before discharge. In compliance to environmental regulations issues like groundwater contamination; aromatics; oil, grease and organic removal, and VOC control have to be addressed during the treatment. A typical wastewater plant in oil and petrochemical industry includes neutralization, coagulation/ flocculation, floatation/ sedimentation/ filtration, clarification and biodegradation (*e.g.*, trickling filter, anaerobic treatment, and aerated lagoon, rotating biological contactor and activated sludge). Finally prior to discharge filtration, ozonation, activated carbon, or chemical treatment may also be required in order to comply with the limits assigned by the environmental regulation authority. The primary intricacy that arises during the treatment of this effluent is the fouling of the membrane because of the presence of various foulants in the feed stream. Therefore, the membrane application here is not only framed by the study on different mechanical parameters' values (*e.g.*, flow rate, dilution etc.) for the feed stream, but also a constant research is going on the fabrication of membrane in order to develop low fouling membrane (Asatekin and Mayes, 2009). Now, what we will do? We will consider individual applications for the oil and petrochemical industry along with an understanding of the insights of membrane technology.

During the exploration of oil (Fig. 2.1) from sea-belt the saline water comes at the surface known as produced water. This contains

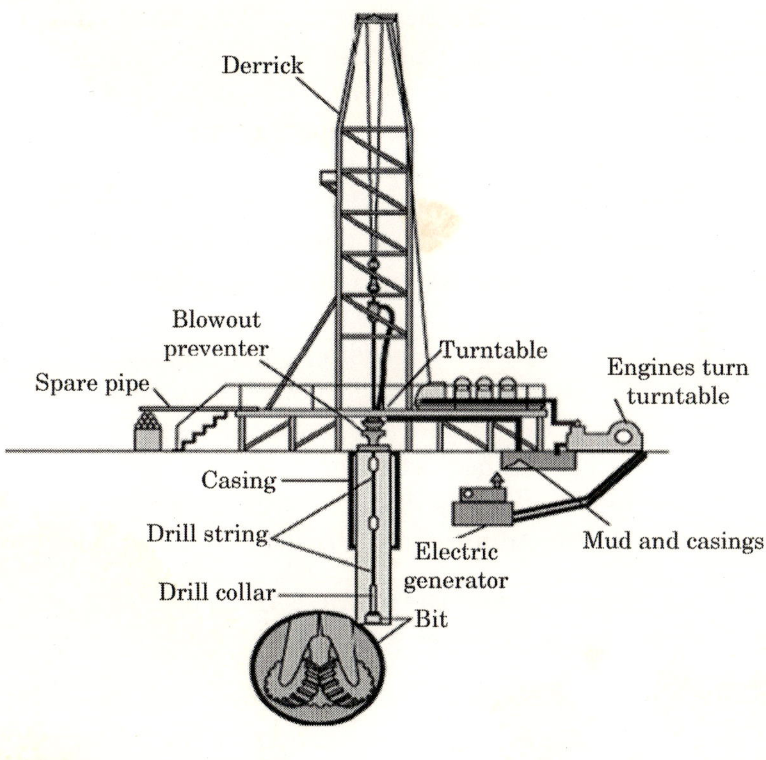

Fig. 2.1: Drilling platform for oil exploration[1]

[1] *Reference*: http://science.howstuffworks.com/environmental/energy/oil-drilling4.htm

salts, heavy metals, oil, and other organics, including polycyclic aromatic hydrocarbons (PAH), phenols, and traces of chemical additives that are used in the drilling operation. However, the composition of this produced water can vary greatly with the location and the life of the well. Further, treatment and disposal of produced water is largely expensive, and needs constant improvement with increased discharge quality. For onshore disposal of produced water or its reuse, produced water needs to be treated to remove oil, toxic substances, and high salt concentrations. While, in off-shore oil wells, discharge into the sea can be possible without the removal of salinity, but essentially all oil and grease (O&G) contaminants have to be removed. Now, using ultrafiltration (UF) the oil and grease can be removed from the water, while the salinity of the system can't be controlled using UF membrane. Therefore, in this case we need reverse osmosis membrane to retain the salts and get clear water. As said earlier that the removal of oil and grease substantiates fouling of the membrane, which is a major issue in case of UF. Now how this can be eradicated? If one can fabricate the surface of the membrane with enhanced hydrophilicity, then the deposition of oil can be restricted. Or in some cases, instead of playing with the nature of the membrane surface, an indigenous fabrication of the membrane setup might help to reduce the possibility of fouling. If we will design the setup in such a way that it has some accessories that creates a sweep over the membrane then the deposited can be removed immediately. Henceforth, the retention of the oil over the surface and thereby the possibility of irreversible fouling could also be eliminated (Putatunda *et al.*, 2015).

Fig. 2.2 shows the image of the modules the authors Putatunda *et al.* (2015) had applied in order to separate oil from water using UF membrane.

In TFMM, the turbine type arrangement makes a sweeping action on the surface of the membrane. The actual flow arrangement is shown in Fig. 2.3 (Putatunda *et al.*, 2014).

Fig. 2.2: Turbine flow membrane module (TFMM) (*Putatunda, S., Sen, D., Bhattacharjee, C., 2015. Emulsified oily-waste water: pretreatment with inorganic salts and treatment using two indigenous membrane modules. RSC Adv. 5, 52676–52686- Reproduced by permission of The Royal Society of Chemistry*)

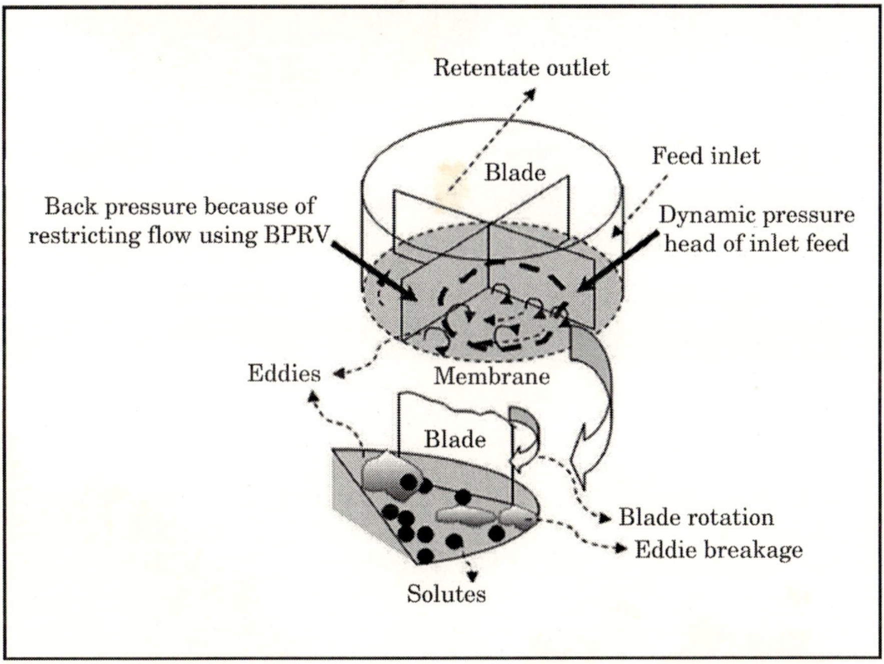

Fig. 2.3: Schematic representation of the feed flow within TFMM (*Putatunda, S., Sen, D., Sarkar, A., Chowdhury, R., Sarkar, D., Bhattacharjee, C., 2014. Two indigenous high sheared membrane modules' performance expatiation for the ultrafiltration of polyethylene glycol. RSC Adv. 4, 6435–6446- Reproduced by permission of The Royal Society of Chemistry*)

However, another major concern with the membrane separation process is the mechanical strength of the membrane. As the fouling increased during the treatment of oily wastewater, hence the process needs to be shifted sometimes to a varying pressure system within the tolerable range of the pressure. However, the technique sometimes fails due to the mechanical failure of the polymeric membrane. Hence, in order to the increase in the strength, the membrane can be inorganic membrane like ceramic membrane, zeolite membrane etc., which gives enough mechanical strength to the membrane to restrict such failures. Fig. 2.4 shows two images of inorganic and polymeric membrane surface fouled by oil deposition. Fig. 2.5 shows the schematic of wastewater treatment for oil-water separation.

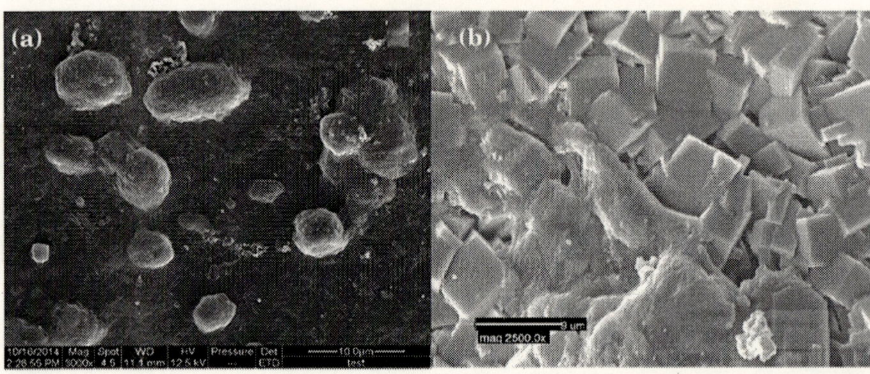

Fig. 2.4: Fouled (a) polytetrafluoroethylene (*(Zhang et al., 2015)- Reproduced by permission of The Royal Society of Chemistry*) (b) zeolite membrane (Zhu *et al.*, 2013)

2.2. MUNICIPAL WASTEWATER

One of the largest issues probably nowadays people are paying attention. If you look at the above Fig. 2.6, there is almost a gradual elevation in the municipal wastewater generation for the last three decades. The questions are "Why" and "How"? The obvious answer to "Why" is because of constantly growing urbanisation. However, answer to "How" is a bit difficult and tricky also. And moreover,

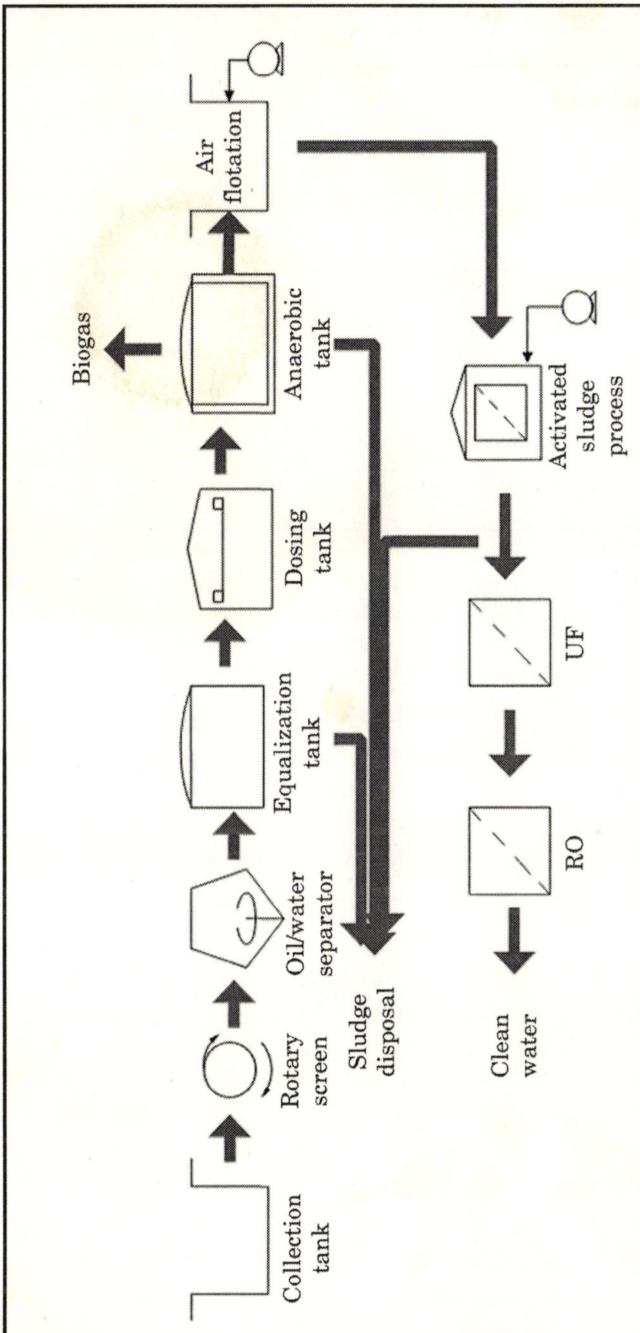

Fig. 2.5: A schematic of the oily wastewater treatment plant

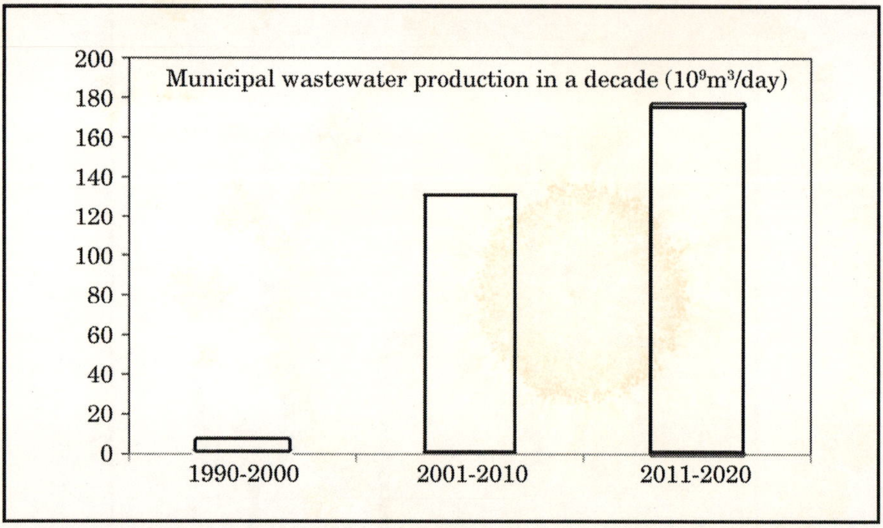

Fig. 2.6: World municipal wastewater production scenario (AQUASTAT from Food and Agricultural Organisation of the United Nations; *Source: http://www.fao.org/nr/water/aquastat/wastewater*)

this is the onset of designing a treatment process in order to mitigate the effect of waste stream on our socioeconomic life. These two questions are interrelated and also the effect of our socioeconomic condition also. So if we will think of the whole process of municipal waste generation and their impact, we will probably have some interdependent issues – first one is impacting the second one and the abundance of the second one might create the first one. If you just look at the Fig. 2.7, it will give you some classical image of such cycle.

In Fig. 2.7 there could be two ways the pollutants will be discharged to the water body. One, which is direct because of the run-off water from locality to the river (shown grey line) and indirect, where the water will be collected in a pit followed by subsequent treatment. The later one is one of the prime intentions. However, because of such direct discharge the aquatic zone gets polluted and the usage of such water needs hospitalisation (black dashed line). Now how the cycle executes? Because of economic

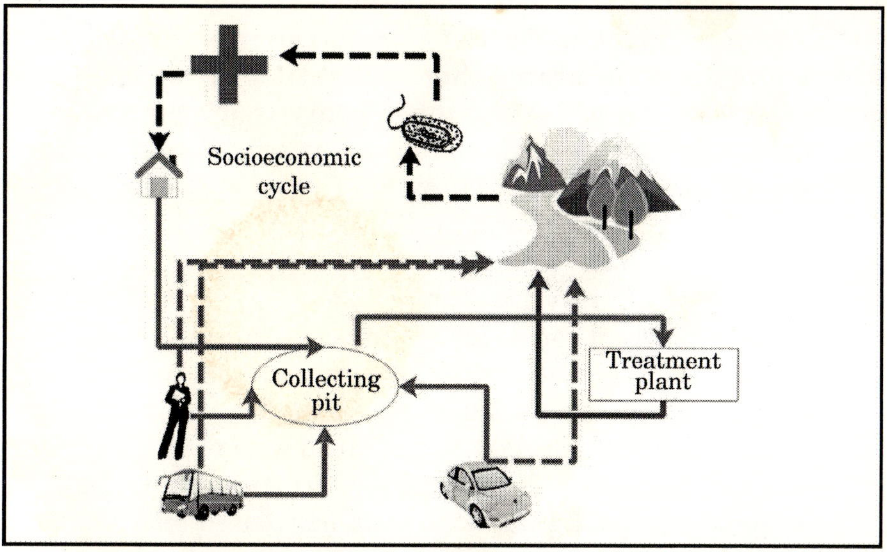

Fig. 2.7: Interrelationship between wastewater generation and socioeconomics

booms and therefore, random urbanisations increase the wastewater load. This results an increase in water borne diseases, which eventually increases the hospitalisation and this will create needs for more urbanisation in terms of infrastructure development towards the updated medical facilities. Therefore, we may call it a cycle, which is continuously growing and cannot be stopped unless we are paying attention to the restriction on the contamination. Adequate treatment of wastewater and a simultaneous supply of sufficient clean water have become a major concern for many communities.

Before, going to an in-depth understanding of wastewater treatment process let us see the parameters on which we may understand the quality of the wastewater. Oxygen demanding substances (after measuring the BOD, COD, DO etc. of water), pathogens and nutrients are the constituents of wastewater, which needs to be measured in order to judge the quality of water. Therefore, the scheme for water treatment should be designed in a way, that we can control the parameters under the permissible value. Membrane might be the best alternatives in order to do so. Before going to understand the application of membrane separation process let us look at the primary treatment that is required to reduce the load on membrane system. Fig. 2.8 shows the process schematics for the treatment of sewerage wastewater and the run-off water. According to the figure we segregated the process in three streams – (a) primary treatment (b) secondary treatment and (c) tertiary treatment. The sewage is initially allowed to pass slowly through large tanks, which are commonly called primary clarifiers or primary sedimentation tanks. The tanks are large enough, where the sludge can settle and materials such as grease and oils can float to the surface so that it will be skimmed off subsequently. The primary treatment produces both an almost homogeneous liquid, which can be treated biologically and a sludge that can be separately processed. Primary clarifiers are usually equipped with mechanically driven scrapers, which continuously drive the collected sludge to a hopper in the base of the tank from where it can be pumped to further sludge treatment process. The clarified water flows on to the secondary treatment. Secondary treatment processes can remove up to 90% of the organic matter in wastewater through biological treatment processes. The two most prominent conventional processes used to achieve secondary treatment are attached growth processes and suspended growth processes.

In attached growth (or fixed film) processes, bacteria, algae, fungi and other microorganisms are allowed grow and multiply on the surface of stone or plastic media. This forms a microbial growth or slime layer (biomass) on the substrate media. Wastewater passes over the substrate aerobically, and the bacteria consume most of the organic matters in the wastewater as their food. Attached growth process comprises of trickling filters, biotowers, and rotating biological contactors. In suspended growth processes, the microbial growth is suspended in an aerated water mixture where the air is continuously introduced in an agitated mixture, to allow effective oxygen transfer. The aerobic bacteria and other microorganisms break down the organic matter in the sewage.

Acclimatization of microorganisms in the wastewater in the form of a suspension is usually done for several hours. Suspended growth process comprises of a variations of activated sludge, oxidation ditches and sequencing batch reactors. After biological treatment, the water is fed to secondary clarifiers where any residual solids and the microorganisms sink to the bottom. These solids are handled separately from the supernatant and fed to the disinfection stage. Tertiary treatment comprises of primary the final treatment to the wastewater in order to reduce its trace toxicity and make it convenient to discharge. Coagulation, filtration, reverse osmosis and nutrients fixation are the steps of treatment.

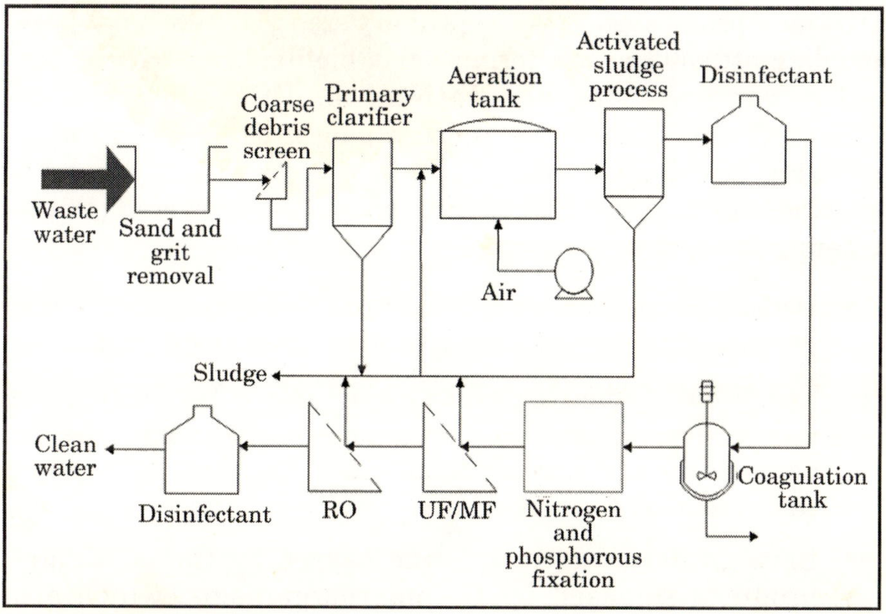

Fig. 2.8: Schematic of municipal wastewater treatment process

At the tertiary treatment process, with the UF process, for a direct ultrafiltration process following the microfiltration the nutrients, compounds contributing to COD will pass through as the permeate stream, while the other pathogens will be retained over the membrane. Now one of the crucial demerits of applying direct membrane process without implementing either of primary or secondary process is the membrane fouling due to the development of biofilm over the membrane. This development will be multiplied as the nutrients are also getting stuck on the surface due to drop in the permeation because of biofouling of the membrane.

The biofouling aggravates, when a membrane stage is used at the outlet of a wastewater treatment plant for disinfection of the effluent either to comply with higher standards or to reuse the treated waste water. Disinfection using membrane has advantages compared to other conventional processes like UV treatment, ozonation or chlorination. One of the major advantages here is no formation of undesired by-products. The adhesion of microbial cells on the membrane surface is the primary step towards membrane biofouling attributing to a formation of biofilm layer. Hence, the biofilm comprises of populations of different types of microorganisms such as bacteria, algae, protozoa and fungi. Biofouling of the membrane during the retention of bacterial separation can have several adverse effects (Nguyen *et al.*, 2012) on membrane systems such as:

- Decline in membrane flux due to the formation of a low permeability biofilm growth over the membrane surface.
- The differential pressure or we may say the TMP across the membrane and therefore, the corresponding feed pressure being needed to increase to maintain the same production rate due to biofilm resistance.
- Biodegradation of membrane caused by the acidic by-products released by the microorganisms, which are concentrated at the membrane surface. For example, cellulose acetate membrane is more susceptible to being biodegraded because of the presence of cellulosic chain.
- Salt passage through membrane will also be increased (as in case of RO membrane) and therefore, the quality of the product water will also be decreased.

- Due to the accumulation of dissolved ions within the biofilm at the membrane surface, the degree of concentration polarization gets increased manifesting a low permeation.
- Energy consumption will be higher due to the requirement of higher pressure to overcome the biofilm resistance and the flux decline.

Fig. 2.9 shows an image of the biofouled membrane because of *S. cerevisiae* and *E. coli* over a 0.2 μm microfiltration ceramic membrane.

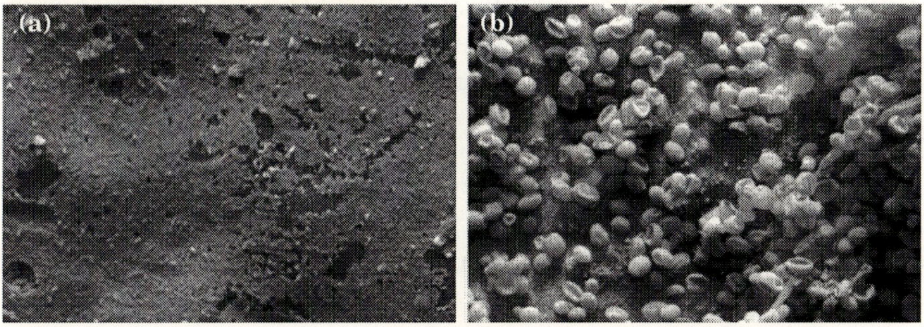

Fig. 2.9: (a) Clean membrane (b) Membrane after biofouling (*Reproduced by permission from MDPI, Basel, Switzerland* (Hassan *et al.*, 2013) *under Creative Commons Attribution license*)

2.3. AGRICULTURAL WASTEWATER

The backbone of a country depends on the strength of their agricultural scenario and hence, the demands for agricultural needs show an astounding growth of the crop market in the global market. Therefore, with the agricultural booms, the wastewater generated from the ground will be in high volume. However, requirements for the agricultural wastewater treatment plant are not set standards as in municipal wastewater. In general the common fertilizers apart from the available fertilizers in the market, cow dung, slurries from animal farms such as dairies and piggeries are spread on land as a fertilizer. Now these materials, with the irrigation water used for agriculture along with other bio and chemical compounds are drained into either the flowing water stream like river or stagnant one like pond, lakes etc. as shown in Fig. 2.10. Now, treating such dirty water from the field is cost intensive and difficult to treat because of the versatility in their toxicity and adverseness. Therefore, a poor country, and especially small villages in that country, can't afford to produce clean drinking water, and water for irrigation purposes. Sometimes farmers use water even from rivers, which are polluted with the sewage or runoff from livestock farms, industry along with other contaminated sources to their cropland. This allows bacteria to be invading into the food, people and animals who live off this, can suffer from serious health issues. Therefore treatment of the water pre and post of the irrigation to the crop field is something much required for having a good agriculture.

Fig. 2.10 shows a picture because of agriculture run-off water disposed to a central flowing water stream. Now, what the constituents of runoff water from agricultural field? Let us make some points on this:

- Agricultural runoff has a load of constituents from soil erosion, feeding operations, grazing, plowing, animal waste, application of pesticides, irrigation water, and fertilizer. It also triggers algae blooms in coastal waters as shown in Fig. 2.11, and produce "dead zones" in the ocean where there will be no oxygen towards the survival of aquatic living beings.
- Agricultural runoff may be the reason for bad taste and odor in drinking water and responsible for contaminating drinking water, well water, and food sources. The pesticides in runoff might be accumulated in fish, and hence be the responsible for contaminating the food chain.

As said before, there is no such treatment protocol designed so far on the treatment of agricultural run-offs. The water coming out of the field might be drained to a wetland and this tail water is

Fig. 2.10: Runoff from farmland near the Santa Maria River drains from ditches to the river (*Source: http://seaotters.com/2012/08/agricultural-runoff-bad-for-us-bad-for-otters/*)

Fig. 2.11: Nitrogen pollution is responsible for the growth of harmful algal bloom (*Reproduced by permission (Grossman, 2016) under Creative Commons Attribution license*)

a rich source of nutrients. Therefore, collection of this tail water and desalination treatment is helpful in recovering the nitrogenous compounds that might be used in fertilizer industry. Now, the treatment is as good as we had discussed in case of municipal water treatment. However, the system becomes loaded as it carries lot of soils as suspended solid with it. In a recent report studied ("Energy-Efficient Process for Using Membrane Technology to Treat and Recycle Agricultural Drainage Water," 2016) the stress is given on a energy efficient process, where prior to filtration operation the solid particles are deposited using gypsum, which is removed through a subsequent operation of membrane separation. Fig. 2.12 shows the scheme on this. In the figure, the last stage comprising of RO provides ultimate purified water coming out as permeate, with no dissolved solid and with permissible limit of BOD and COD.

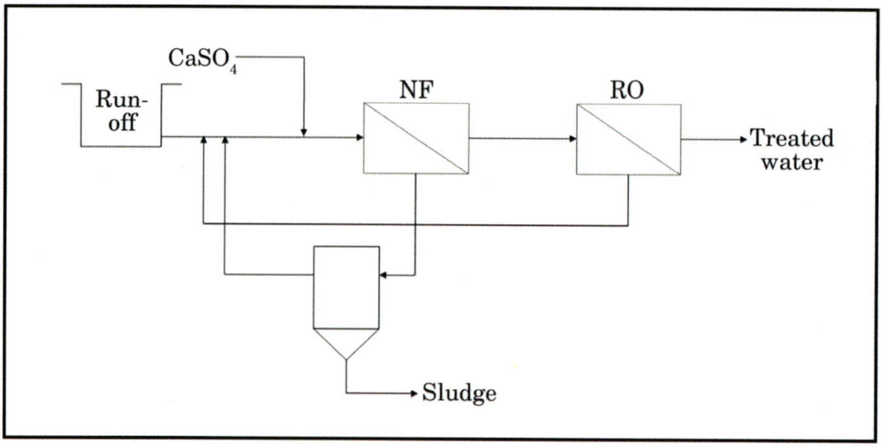

Fig. 2.12: Treatment scheme for agriculture run-off

Because of the variability in the constituents of agricultural run-offs, the selection of membrane is one of the important aspects. As per the previous images it could be seen that membrane process could be intervened early for treating the wastewater. Now the question comes, how far the method is feasible in order to a prolonged use of the membrane? Secondly, if still we can implement the process because of its simplicity, what should be the type of membrane material we should use in the treatment? Secondly, what should be the type of the membrane – whether it is an inorganic membrane or an organic membrane? As said before, for this kind of system, like municipal wastewater treatment, cellulose membrane must be avoided. However, this membrane is hydrophilic and a good option for the wastewater treatment. Therefore, in order to compensate the process, *i.e.,* in order to use cellulose acetate membrane, one must reduce the bacterial content and the dissolved oxygen as well for the system. This manifests a less damage to the cellulose acetate membrane and hence, prolonged used could be possible. In a study, addition of strong oxidizing procedure, like chlorination might help to reduce the contaminants and hence, the membrane can be preserved (Shuval, 2012). Further, as also said before, inorganic membrane has more mechanical strength and hence, mechanical failures can be restricted with the ceramic membrane (inorganic membrane having high hydrophilicity).

2.4. CHAPTER IN BRIEF

- Oily wastewater treatment and oil exploration industry.
- Concept of high sheared membrane to reduce the fouling of the membrane.
- Municipal wastewater treatment and green discharge.
- Agricultural wastewater treatment.

Chapter 3

APPLICATION OF MEMBRANE IN BIOPROCESS

Alcohol – one of the most essential products, probably we have in today's world, which is having a wide applications' area at all levels of our society. Food and beverage industry, fuel industry, solvent manufacturing industry, cosmetic industry – everywhere the alcohols are one the value added chemicals that are use or produced. Now, how we can produce this alcohols. Let us confined us to food and beverage sector, where ethyl alcohols are produce and can be considered as one of the value added products that will contribute a large to generate revenue. So, what the people are doing over there? The key step to produce ethanol is the fermentation process of monosaccharides or we may call it sugars using any organisms like *Yeast*. Further, to produce such monosaccharides people are using carbohydrate enriched natural resources like grains, corn cobs, etc. They are hydrolyzing the cellulosic part to produce monosaccharides and these monosaccharides are being used as a feedstock for the fermentation process to get an ethanolic solution of water. This solution is fed to a distillation column to separate water from the ethanol and obtain ethanol as a product. So in brief, we should use renewable resources to produce carbohydrates, which on subsequent anaerobic fermentation produce ethanol followed by a separation of ethanol from water through distillation. This is what we call

'*Bioprocess*' and the engineering of the process is called '*Bioprocess Engineering*'. Therefore, an engineered process through which one can convert renewable resources into value added product is called '*Bioprocess*'. In this chapter, we will discuss a case study, which is going to give us some thorough idea on different chemical processes involved and how they can be engineered. Moreover, we will try to understand here the application of membrane based advanced separation scheme in order to replace the existing conventional process and how we can be benefitted with the quality of the product at low expenses.

3.1. PRODUCTION OF BIOBUTANOL FROM RENEWABLE RESOURCES - A CASE STUDY

Butanol (C_4H_9OH) is an alcohol containing four carbon atoms, whose boiling point is 117°C under atmospheric condition with a vapour pressure of 7 mm of Hg at 25°C. On the contrary, the vapour pressure of ethanol (C_2H_5OH) is 45 mm of Hg at this condition, which is almost seven times of butanol. Apart from several usages of these alcohols, the common use for both of them as fuel. Now the question is – what is fuel all about? Something, which provides energy after burning, called fuel. Like we all know that for running a car we need engine which will run by gasoline and through burning it releases energy that will run the car. So, *the engine should be a device which can convert heat energy of fuel into mechanical energy*.

During the production of biobutanol, the biomasses collected are segregated according to various parameters observed there. Based on the origin of a tree, the plants are of two types – one is called 'hardwood' and the other is called 'softwood'. Now, hardwood is originated from the seeds having cover wrapping it (angiosperms), while softwood is originated from the uncovered seed falling on the ground (gymnosperms). Now a plant material composed of three layers as we are having. The outer layer is lignin, second layer is hemicellulosic and the inner core layer is cellulosic. Fig. 3.1 shows the network structure of a plant.

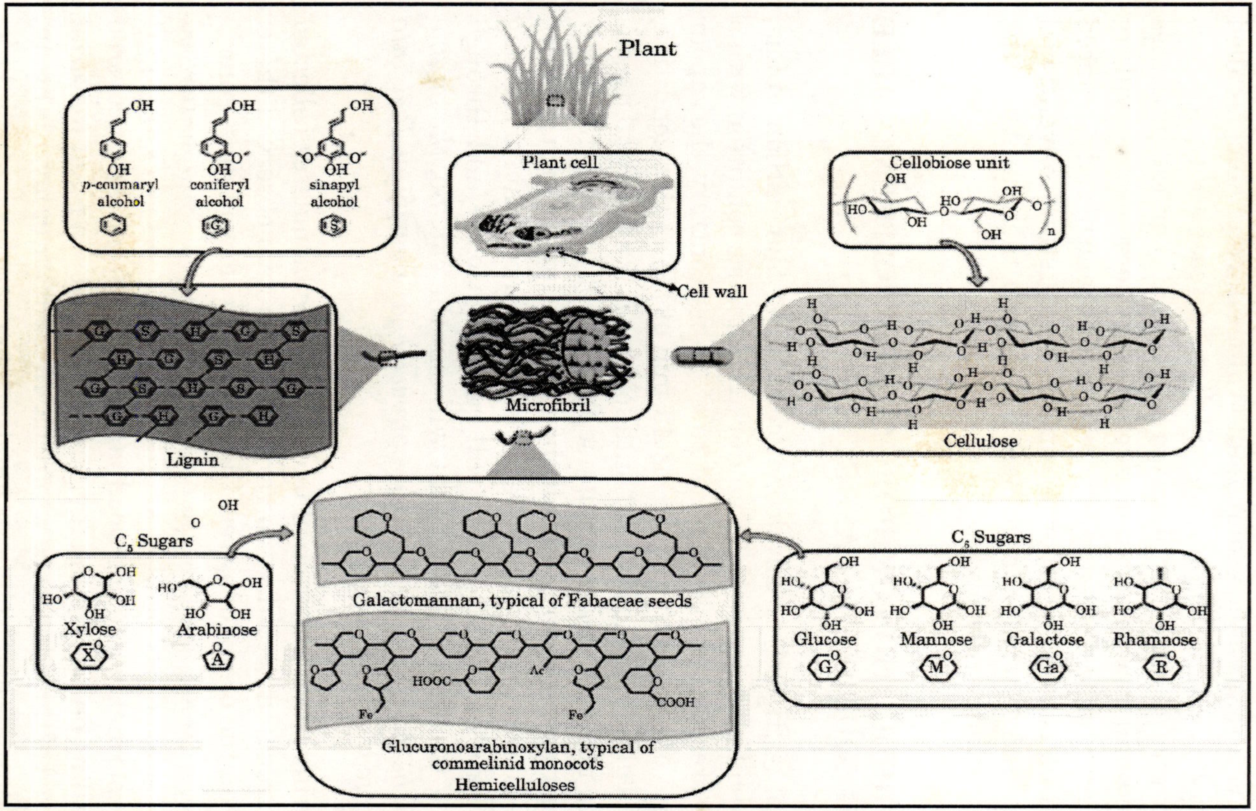

Fig. 3.1: Lignocellulosic network structure of biomass (adapted from (Isikgor and Becer, 2015) *http://pubs.rsc.org/en/content/articlehtml/2015/py/c5py00263j*)

The complex network structure of lignocellulosic biomass restricts its fractionation, especially when during the production of butanol the accessibility of cellulose becomes limited because of hemicelluloses and lignin wrapping on it. Therefore, pretreatment is required to increase the accessibility of cellulose through disrupting the lignin and hemicellulose network. A great deal of different pretreatment schemes and their pros & cons have been discussed elsewhere (Haldar *et al.*, 2016). In this chapter, we shall discuss the post pretreatment schemes, where membrane separation might be one of the crucial unit operations that are expected to enhance the commercialisation of biobutanol. The process schematic towards the production of biobutanol is given in Fig. 3.2.

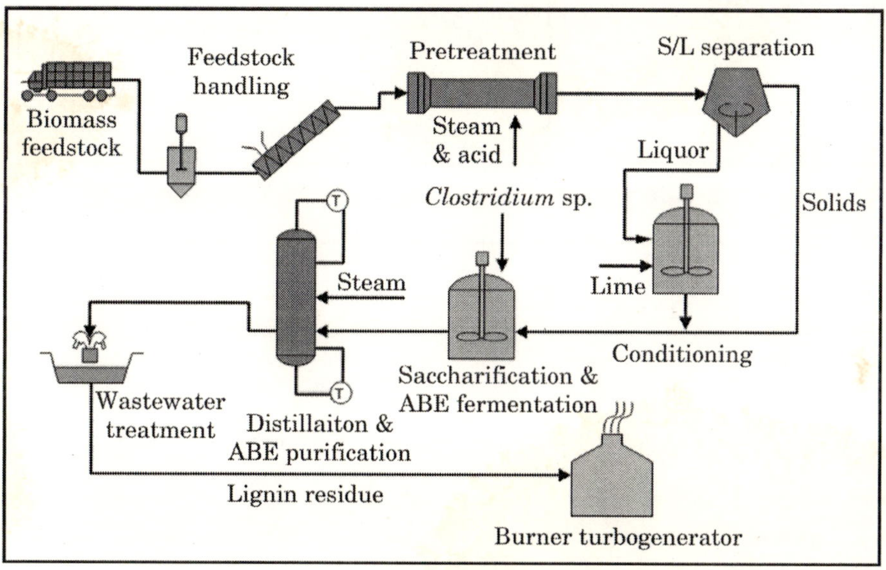

Fig. 3.2: ABE fermentation process

In the figure, for the hydrolysis of cellulose and hemicellulose, the mostly used process is acid hydrolysis, where the 1-4 linkages (Fig. 3.3) will be cleaved and releasing monosaccharides. However, during process of monosaccharides production with acid hydrolysis, the monosaccharides will be hydrolysed (Fig. 3.4) further after releasing three water molecules and produce furan compounds. These furan compounds act as an inhibitor to the subsequent fermentation process and results less yield. Further, the fermentation product contains acetone, butanol and ethanol. The separation of these compounds is critical and high cost will be involved in getting purified form of butanol through conventional process, which can be used as a fuel for complete replacement of gasoline.

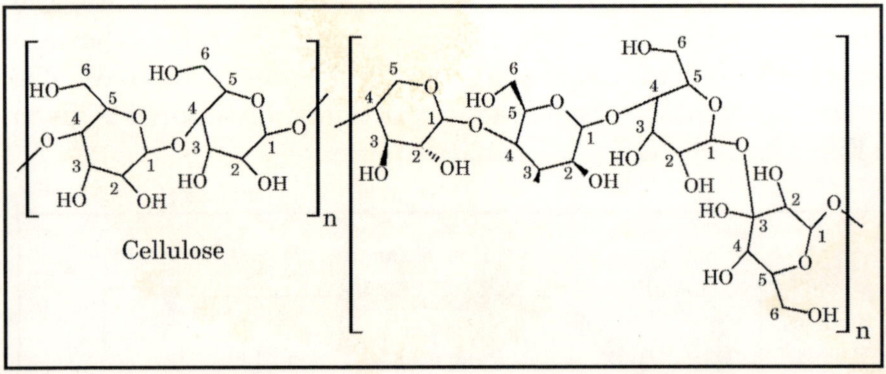

Fig. 3.3: Structure of cellulose and hemicellulose

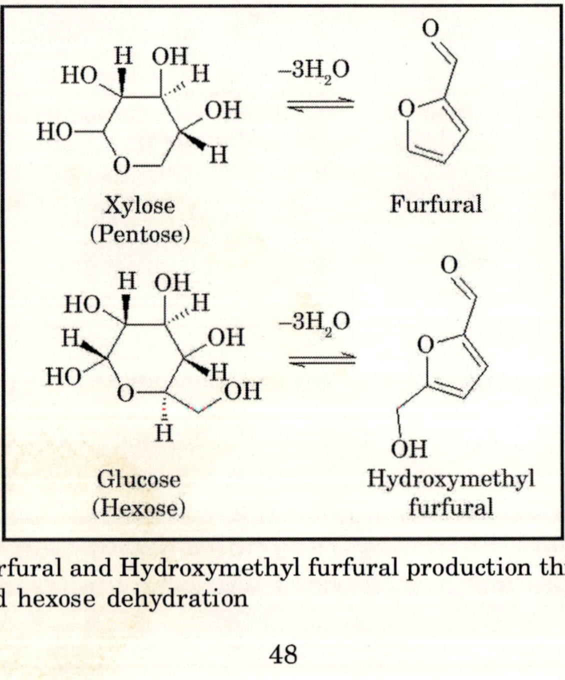

Fig. 3.4: Furfural and Hydroxymethyl furfural production through pentose and hexose dehydration

Production of furan compounds can be restricted after using enzymes to hydrolyse cellulose and hemicellulose. Here, the substrate cellulose and hemicellulose will be cleaved in presence of enzymes. After doing such enzymatic reaction, the enzymes are getting separated using membrane of higher MWCO from the product and the reaction mixture composed of oligomers along with the enzyme is again fed back to the reactor. The arrangement is called enzymatic membrane reactor on we will do some elaborations in the subsequent sections.

3.1.1. Enzymatic Membrane Reactor

Membrane reactor (MR) consists of a semipermeable membrane, which allows the free passage of the product molecules but contains the enzyme molecules. However, one of the critical concerns in membrane reactor development is the choice of the membrane as it takes high molecular weight protein molecules on its surface and must ensure the reflection of the proteins. The general choice for a membrane reactor is hollow-fibre membrane module containing hundreds of thin tubular fibres of having a diameter of about 200 mm and a membrane thickness of about 50 mm. The details of the membrane reactor will be discussed later. What we will say here is how the membrane could be used here as a separating medium after producing monosaccharides from polysaccharides. Fig. 3.5 shows a schematic arrangement of the membrane reactor, where enzymes are being separated out from the reaction mixture.

Now let me explain the process in details. We have a batch reactor, where we have the pretreated biomass, consists of cellulose, hemicellulose and lignin primarily. In the system, we have dosed enzymes to dissociate the polymers into their respective monosaccharides. Now, before we are employing any membrane to the process, we should know what should be the molecular weight distribution in the reaction mixture. Apart from monosaccharides (C6 and C5), acetic acids the other components are oligomers and having high molecular weight more than 1000

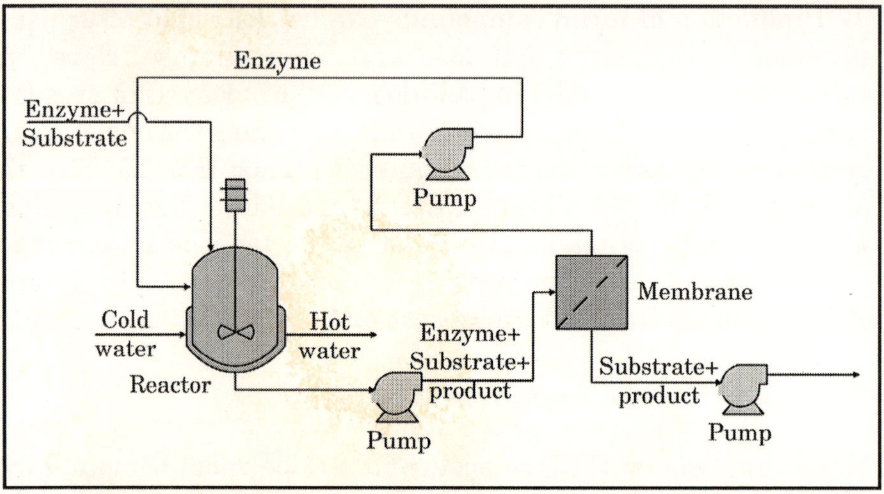

Fig. 3.5: Enzymatic membrane reactor arrangement

Da[1]. "One of the primary advantages with the membrane reactor is using it as a separating medium preventing any sort of morphological or chemical changes with the components." – Now, what does it mean? In this case, initially the enzyme is being separated out from the reaction mixture using an ultrafiltration membrane of adequate molecular weight cut-off. Permeate from the membrane consists of carbohydrates of varied molecular weight will be given to a nanofiltration membrane to collect the monosaccharides into a fermentor and the uncleaved oligosaccharides will again be fed back to the reactor. Fig. 3.6 shows the scheme. So we can understand that without too much complexity we can design an effective scheme to separate a solution mixture. Now, what is the disadvantage of the procedure? The most critical consideration is the membrane fouling as discussed before. Especially, with the proteins and carbohydrates, there is a great deal of adsorption of such organic molecules on the surface (Absolom *et al.*, 1987; Sen *et al.*, 2011b). This will reduce the membrane throughput and also restricts the regaining of flux

[1] A unit used in expressing the molecular weight of proteins, equivalent to atomic mass unit. It is an expression of weight per mole.

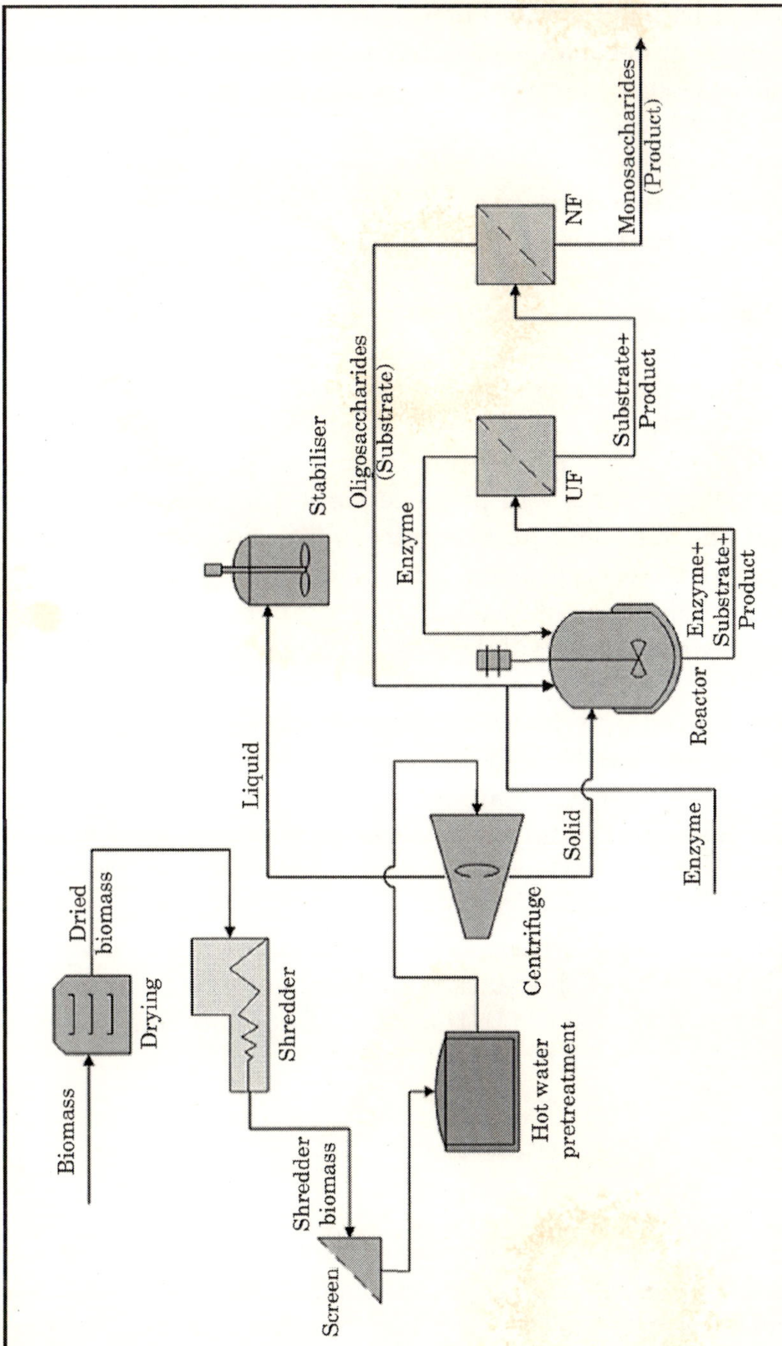

Fig 3.6: Arrangement for the synthesis and purification of monosaccharides from biomass

because of irreversible fouling occurred over the membrane. The mathematical details for the enzymatic membrane reactor have been discussed in chapter 5, where the discussion on membrane reactor is elaborated.

3.1.2. Acetone-Butanol-Ethanol (ABE) Separation: Application of Pervaporation

The reaction product comprises of monosaccharides after being separated out the oligosaccharides present in the post-hydrolysate will be given to the fermenter. The conversion of the fed monosaccharides to the fermenter is converted into alcohols (C_2 and C_4) along with acetone using a bacterial strain called *Clostridium sp.* This will produce three components – acetone, butanol and ethanol and the process is called acetone-butanol-ethanol (ABE) fermentation. You all know, that when we need to separate such a mixed system with components of high vapour pressure, we need distillation to be there as a unit operation. Depending on the boiling point difference, the mixed system can be fractionated and different components can be separated out from the operation. Now, if the components will be of high vapour pressure, the energy consumption will be least due to reduced reboiler load. But, what is the crucial point without which we even cannot think of the applicability of distillation? The relative volatility will guide us to understand the possibility of using distillation for the system. The definition of relative volatility ($á_{AB}$) is the ratio of the mass/mole fraction of one component in two phases with respect to the ratio of another component (eq. 3.1).

$$\alpha_{AB} = \frac{\left(P_A/x_A\right)}{\left(P_D/x_B\right)} = \frac{\left(y_A/x_A\right)}{\left(y_D/x_B\right)} = \frac{y_A x_B}{y_B x_A} = \frac{y_A(1-x_A)}{(1-y_A)x_A} \tag{3.1}$$

Certain combination or proportions of alcohols and water restricts the onset of distillation. That is why, 80% solution of ethanol cannot be further distilled to have more purification of ethanol using distillation. Now let us come to the utility of membrane in such cases, where distillation might have a failure. Pervaporation is a membrane separation process, where a membrane is sandwiched between the feed mixture (mixture, where the intended component is supposed to be present at high concentration) and sweep (stream, where the component intended to be separated presents at low concentration). Now here we create a concentration driving force, where because of the above arrangement said, we have the separable component to be moved on the other side of the membrane. Fig. 3.7 shows a pervaporation schematic.

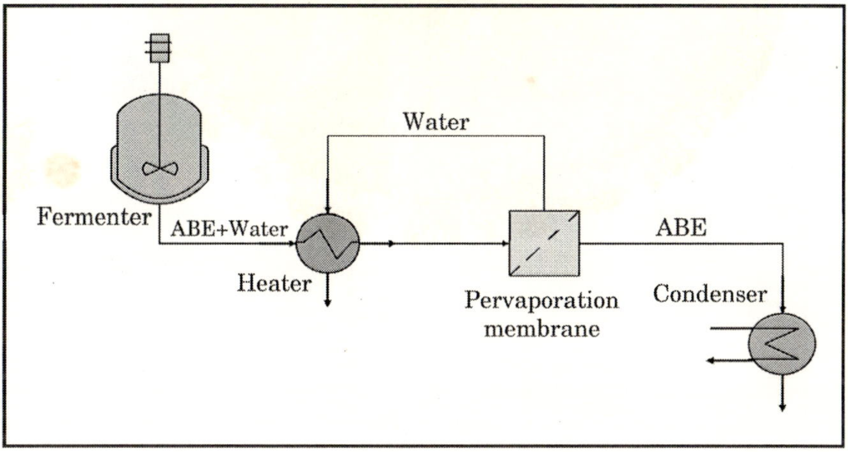

Fig. 3.7: Arrangement of pervaporation set-up

In the figure, the inlet stream is made hot to increase the vapour pressure followed by conversion of state to vapour one. On the permeate side, the partial pressure of the vapour is reduced due to the condensation and the energy released will be reutilised for supplying the energy at the inlet. Sometimes, in order to make the process energy efficient, instead of condensing the vapour at the permeate side a "sweep" stream is used. So based on these three operational aspects, pervaporation will be of three modalities – (a) vacuum pervaporation (Fig. 3.8a) (b) sweep gas pervaporation (Fig. 3.8b) and (c) thermopervaporation

(Fig. 3.8c). Now, what are the fundamentals of separation of butanol, our fuel, from ABE mixture? Before, we are going into the detailing of the separation procedure, let us come to a discussion to understand the requirements for pervaporation.

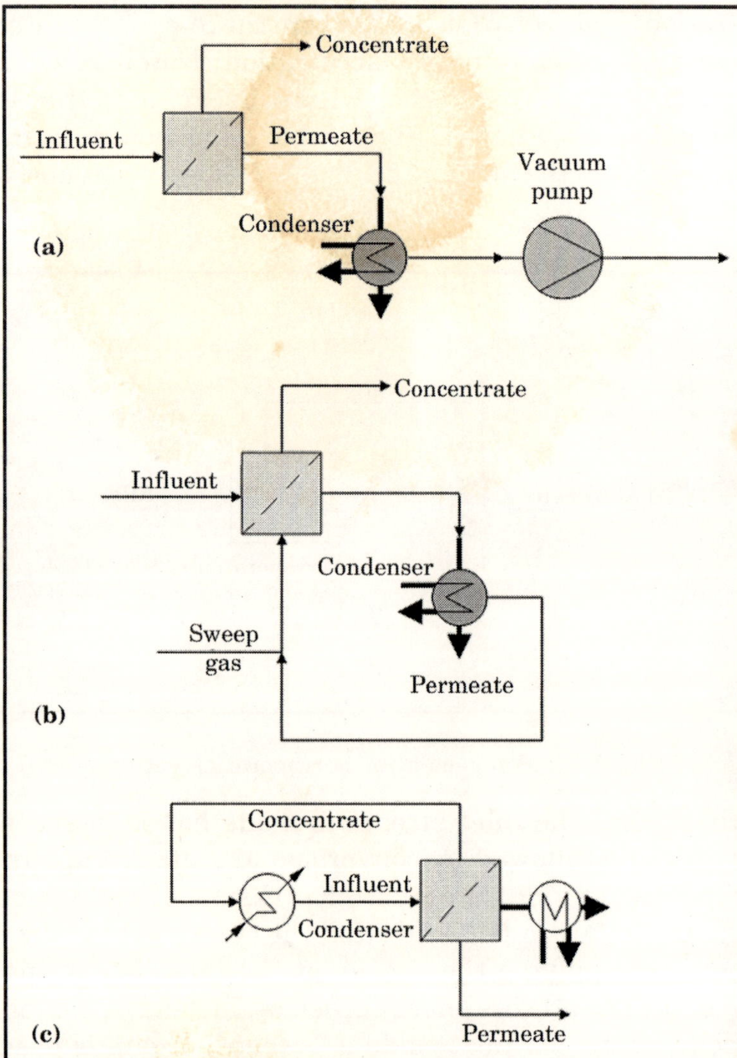

Fig. 3.8: Different arrangements of pervaporation (a) vacuum pervaporation (b) sweep gas pervaporation (c) thermopervaporation

The prime most consideration in any membrane separation process is the membrane – based on its characteristics or nature, pervaporation can be performed.

Pervaporation membranes are typical composites membrane. The first layer of the membrane on the feed side is polymeric and porous. This layer is coated with a second polymer – called the "active" or "permselective" layer, which is engineered to preferentially absorb the chemical species of interest. Varying the thickness of the permselective layer the pervaporation characteristics can be controlled. In general the transport of the species across the membrane can be described by three sequential processes. These are preferential sorption of the desired species on the polymeric matrix of the membrane, diffusion through the membrane and finally, desorption on the permeate side. Now as said before, sorption will depend on the nature of the membrane used. If it is hydrophilic membrane, the sorption will be for water and on the contrary, if it is hydrophobic one, the sorption is for apolar components. In general the porous polyester support casts either a polyacrylonitrile (PAN) or polysulfone along with an almost 0.1 μm cross-linked polyacrylate, poly-vinyl alcohol (PVA) or any other hydrophilic surface ("Pervaporation developments", 1992). Typically, pervaporation separation unit occupies between ten and one hundred m^2 of membrane area must be packaged into units called membrane modules. Flat-sheet and spiral-wound modules are commonly used. Spiral wound configuration provides a high membrane surface area per module and operates for relatively high feed flow rates, common for pervaporation. The primary reasons for employing pervaporation set-up for the ABE fermentation is its continuous implementation with the fermentation process. The butanol produced during the fermentation results an increase in the fluidity of the cell that disrupts the control mechanism of the cell membrane during the transport of the extracellular and intracellular components across the membrane. Therefore, butanol separation from the broth becomes a critical consideration during its production through ABE fermentation. Apart from increasing hydrophobicity of membrane (as butanol is hydrophobic in nature), inclusion of ionic liquid that selectively transports butanol within the porosity of the membrane could be an advantageous process. The concept comes from supported liquid membrane technology, where the membrane pores are being filled with the ionic liquid acting as the carrier. However, the main disadvantage with the process is the leakage of the liquid and ultimately, the separation becomes hindered. In one the literatures (Plaza *et al.*, 2013), an ionic liquid (IL) 1-butyl-3-methylimidazolium hexafluorophosphate ([bmin][PF_6]) was gelated into the porosity of polytetrafluoroethylene (PTFE) hollow fibre membrane. The IL-gelator mixture was prepared using 0.4% (w/w) 12-hydroxystearic acid (gelator) to avoid moderate cross-linking.

Initially the membrane housing was kept at 80°C with an electric tape. 10 ml IL-gelator mixture was circulated into the lumen-side using a peristaltic pump and simultaneously a vacuum was maintained on the shell side by means of a vacuum pump until a constant permeation flow was observed. After the completion of the process nitrogen was simultaneously introduced in the lumen and shell side in order to clean all the excess gelator-IL mixture. In order to achieve the solidity and stability of the membrane, the temperature of the module was then decreased from 80°C to 20°C at a rate of 10°C/min. It is seen that the selectivity, defined as ratio of the mass of component to be separated and mass of the other component in the mixture, of butanol over ethanol is around 1.3 to 2.1 and over acetone it is around 1.1 to 1.2. However, it is a very specific figure referred here, while the system or the separation could also be adjusted through a variation in the sweep.

3.2. PERVAPORATION MATHEMATICS

Before going into the mathematical insights of pervaporation process, let us discuss on a topic of thermodynamics – called "non-equilibrium thermodynamics". One of the mostly discussed and I must say, the more realized topic in thermodynamics is the state of equilibrium. Now, let us take three systems. One, which is not moving – that means the displacement of the system with time is zero. Here, the displacement means it can be physical movement or the transfer of state. Once, we have that time independent system, we can call it a system belong to its "steady state". If I am saying now, the system is in equilibrium, whether I am pointing out its "steady state" condition. Let us go back to our school days, when we came to learn the principle of Le Chatelier stating that "*If a dynamic equilibrium is disturbed by changing the conditions, the position of equilibrium moves to counteract the change*". This is what the primitive law of equilibrium, where we can differentiate the steady state and the equilibrium. The example is with the reversible reaction, where the system continuously moves from reactant to product and *vice versa*. Thus the transfer of state is varying with time, while the net change in the system is somehow managed to get the product yield same *i.e.*, under equilibrium.

Now the question comes, how can we prepare two similar systems are in equilibrium? The system under equilibrium ones does not change even in the very long waiting time. Say for example, the total number of particles, or the total energy. Let us consider a box containing a fluid (gas or liquid). Then with specifying the amount of a one component in the fluid (*i.e.*, the number of molecules), the volume, and the energy, is sufficient to completely determine the equilibrium state. This means that if these quantities, amount, volume and energy are the same for two similar systems, and then all other macroscopic or statistically averaged quantities will also be the same. More specifically, the thermodynamic properties of an isolated system can be completely determined by the volume and the values of the conserved quantities (energy and amount) of the system. These are therefore known as thermodynamic variables. However, in reality, the existence of the isolated thermodynamic system might be difficult. The system might be changed from its steady state with a small perturbation in the system and even long waiting time to come to equilibrium is a bit difficult. Thus we may say this dynamicity in the system is its *non-equilibrium* state. Let us elaborate the topic with an example shown in Fig. 3.9.

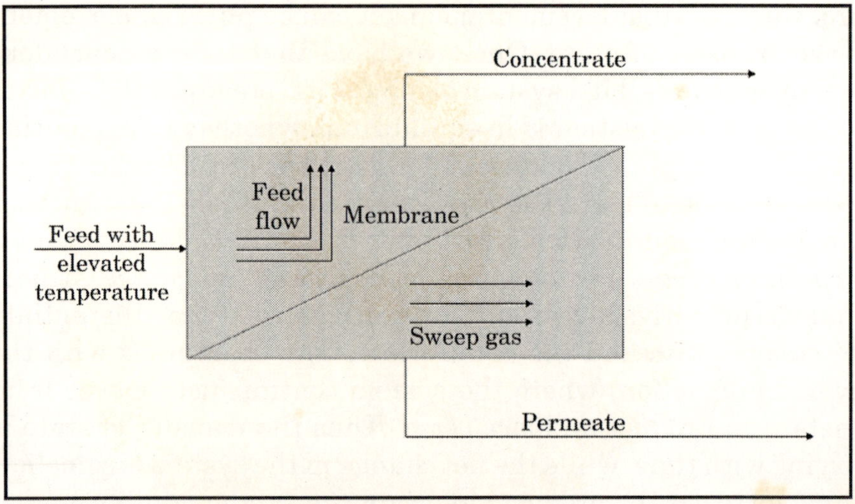

Fig. 3.9: Pervaporation unit (sweep gas pervaporation)

In the figure, the feed with at elevated temperature was fed to the membrane unit and on the permeate side, we have the sweep stream. Depending on the partial pressure of the separable component, the vapour component passes through the membrane on the permeate side, which is condensed. Now coming back to the concept of steady state the accumulation on the membrane can be neglected, which eliminates the transient behaviour of the system. But what will be about the equilibrium? We can't say the system will be in equilibrium even after waiting for long. We will get the dilute concentrate over the time ensuring good separation efficiency. But, we can say the system always tries to stabilize its movement, a tendency towards the equilibrium. Therefore, the dynamics of the system are dissipative, where an injection of energy will be on one boundary (with feed here), energy currents within the system (feed side to permeate side), sometimes a transformation between different forms of energy (*e.g.*, mechanical or chemical to heat), and then the subtraction of energy to maintain the steady state (on the permeate side). The question is what should be the thermodynamics model in this case, where we have a true thermodynamic vapour-liquid state. The answer is, this stream of thermodynamics is "*non-equilibrium thermodynamics*".

3.2.1. Non-equilibrium Thermodynamics and Membrane Transport

To understand the non-equilibrium thermodynamics, let us concentrate on the second law of the thermodynamics. It says that the spontaneity for a change happens, when the entropy *i.e.,* the disorder of the system increases. Now let us consider a diffusion process, where the molecules are moving on the other side of a permeable barrier because of the concentration gradient across the barrier. Now see the following figure where the gaseous component is diffusing through the membrane.

In the Fig. 3.10, one side of the membrane consists of 8 numbers of molecules at t=0. As the process goes on the two compartments are having equal number of molecules on both the sides. Now entropy is defined as the measure of disorder of the macrosystem and the disorder comes because of different possible arrangements on both the sides of the membrane. Below table shows the number of possible arrangements at different time stamp. The table shows that there is an increase in the possible arrangement once the system approaches towards the steady state. According to the definition of entropy, this is the measurement saying the direction of the change. However, it is not giving us any rate consideration, which is obvious with the non-equilibrium study. This was primarily addressed by Prof. Lars Onsager in 1931. According to him, the rate of entropy decreases during a spontaneous change in the system, which was later modified by Prof. Ilya Prigogine (Attard, 2009). According to him, there is an increase in the rate of entropy production with the spontaneous changes of the system. What does it mean at all? Rate of increase in entropy means with time different combinations of the pattern are also increased. Table 3.1 says about the number of possible patterns.

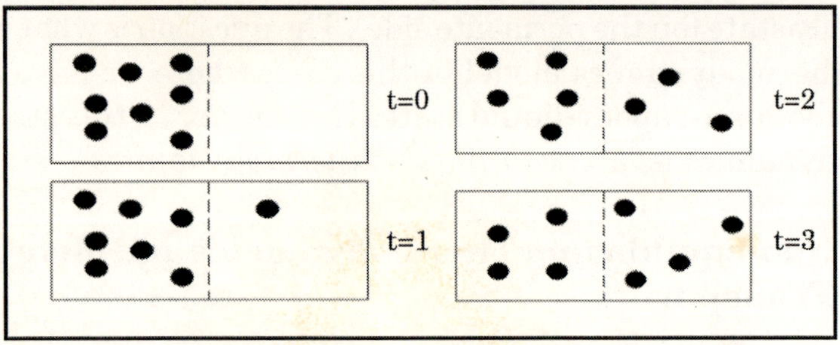

Fig. 3.10: Diffusion of gaseous component across the membrane at different interval of time (t)

This shows the production of new product and obviously this shows the spontaneity. For example in a chemical reaction, the production of a product says about the new patterns and hence, the number of pattern increases. Now, what we will say on pervaporation?

Table 3.1: Number of different possible arrangements at different time scale

Time (t)	*Number of molecules shifted (ns)*	*Number of possible arrangements within the chamber* $\left(\frac{8!}{n_s!(8-n_s)!}\right)$
0	0	1
1	1	8
2	3	56
3	4	70

According to Lars Onsager's statistical theory of irreversible processes one will be able to relate measurable quantities (transport coefficients, thermodynamic derivatives) to the results of experimental measurements. This theory is linear and limited in validity to a vicinity of equilibrium. Now what Onsager said is the flow is somehow the effect of forces and the flux can be given by the terms of the forces correlated by a coefficient called "transport coefficient" or "phenomenological coefficient". According to them the linear relationship for the conjugate flux[2] is given by

$\vec{J}_i = \sum_{j=1}^{n} L_{ij}\vec{X}_j$. Now, the energy of dissipation is synonymous to entropy production. Let us go with the concept through an example. In an enclosed gas chamber the gas is under pressurised condition. Now suddenly the pressure is released and some mechanical work has been performed. Here, the dissipation happens, where the internal energy of the system converts to mechanical energy and the entropy of the system produced, which is explained by the second law of thermodynamics. Therefore, the energy dissipation, given by $\frac{\partial^2 E}{\partial V \partial t} = \sigma = \sum_{i=1}^{n} \vec{X}_i \vec{J}_i = \sum_{i=1}^{n} \vec{X}_i \left(\sum_{j=1}^{n} L_{ij} \vec{X}_j \right)$ (Dickel, 1984). Therefore, from the equation it can be said that the entropy production is proportional or in the order of square of the forces. This gives an idea that if one force is having negative impact compared to the other one the net effect of the entropy production rate will be negative. Let us take one example on that. Say a mixing of binary gases is happening because of the temperature gradient. See Fig. 3.11.

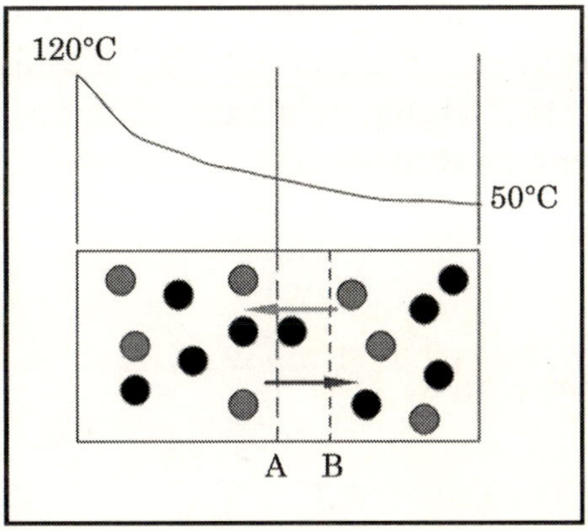

Fig. 3.11: Expansion of the gas and mixing because of temperature gradient

---

[2] Conjugate variables refer to the variables are those that are interchangeable. Like temperature and entropy or pressure and volume

Here, if we go on with the analysis of the system, according to the Onsager's theory, there are two forces that initiate fluxes. One, the temperature gradient expands the gas chamber at higher temperature and pushes the piston from A position to B position. This initiates the mixing due to diffusion of the two components on either side of the piston. So, one force is "T and the second one is "c. Now, the mass flux is related to the concentration differences, or we may say the differences in chemical potential, which is non-determined term. Therefore, "c and related mass flux is called conjugate flux. On the contrary, the temperature gradient and mass flux are in non-conjugate relationship. According to the theory, the conjugate coefficients are on-diagonal and the other coefficients are off-diagonal. Therefore, the mass flux can be written as,

$$\vec{J}_1 = L_{11}\vec{X}_1 + L_{12}\vec{X}_2 = L_{11}\Delta c + L_{12}\Delta T \tag{3.2}$$

$$\vec{J}_2 = L_{21}\vec{X}_1 + L_{22}\vec{X}_2 = L_{21}\Delta c + L_{22}\Delta T \tag{3.3}$$

$\vec{J}_1$ and $\vec{J}_2$ are mass and heat fluxes respectively. In this case, ΔT is negative w.r.t Δc. This yields the rate of change of entropy negative, as suggested by Onsager. Now in equation 3.2 the flux can be varied according to the free force Δc, when ΔT is constant. Therefore, the rate of change of entropy is differentiable w.r.t Δc and according to Prigogine's minimal entropy theory the differentiation can be written as,

$$\frac{\partial \sigma}{\partial(\Delta c)} = 0 \tag{3.4}$$

Again, $\Delta\dot{S}$ can be given by,

$$\begin{aligned}\sigma = \sum_{i=1}^{2} \vec{X}_i\vec{J}_i &= \vec{X}_1\vec{J}_1 + \vec{X}_2\vec{J}_2 \\ &= (L_{11}\Delta c + L_{12}\Delta T)\Delta c + (L_{21}\Delta c + L_{22}\Delta T)\Delta T \\ &= L_{11}(\Delta c)^2 + L_{22}(\Delta T)^2 + (L_{12} + L_{21})\Delta c\Delta T\end{aligned} \tag{3.5}$$

Therefore,

$$\frac{\partial \sigma}{\partial(\Delta c)} = 2L_{11}\Delta c + (L_{12} + L_{21})\Delta T = 0 \tag{3.6}$$

Again according to Onsager's assumption L_{12}=L_{21}. With this assumption, equation 3.6 gives,

$$L_{12}=L_{21} \tag{3.7}$$

$$L_{11}\Delta c + L_{12}\Delta T = \vec{J}_1 = 0 \tag{3.8}$$

Therefore, it might be inferred that with the minimum entropy production, the diffusional flux vanishes and the system will be at equilibrium. One can say that both Onsager and Prigogine are complementing each other. Onsager enlightens minimum dissipation rate, which ensures decrease in the entropy production rate. On the other hand, Prigogine says about the minimal entropy production rate, which in turn says the increase in the rate of entropy production if moving on either side from the minimal point.

3.2.2. Non-equilibrium Thermodynamics and Pervaporation

Any membrane separation process generally executes at isothermal condition, where the lost work or dissipation rate of energy can be expressed as the rate of entropy production irreversibly (equation 3.9).

$$T\sigma = \Delta\dot{H} - \Delta\dot{G} \tag{3.9}$$

Now, the membrane process is open isothermal system and hence, $\Delta\dot{H} = 0$. Therefore, equation 3.9 can be written as (where entropy change is considered across unit area of the membrane),

$$T\sigma = -\Delta\dot{G} = -\sum_k N_k \Delta\mu_k = -\sum_k \left(J_k^D + x_k N\right)\Delta\mu_k \tag{3.10}$$

For continuous changes equation 3.10 can be converted to $T\sigma = -\sum_k \left(J_k^D + x_k N\right) d\dot{\mu}_k$.

Now according to Gibb's Duhem equation for isothermal process one can say,

$$\sum_k x_k d\mu_k = -SdT + VdP = VdP \text{ (dt=0 for isothermal process)} \quad (3.11)$$

Therefore, $$T\sigma = -\sum_k \left(J_k^D + x_k N\right) d\mu_k = -\sum_k J_k^D d\mu_k - NVdP \quad (3.12)$$

Now we know that the molar free energy change is given by $\Delta\dot{\mu}_k = RT\Delta(\ln f_k)$, where f_k is the fugacity. Therefore, equation 3.12 can be written as,

$$\begin{aligned} T\sigma &= -\sum_k J_k^D RT\Delta(\ln f_k) - NV\Delta P \\ &= -\sum_k J_k^D RT\left(\ln f_k^r - \ln f_k^p\right) - NV\Delta P \end{aligned} \quad (3.13)$$

(Here 'r' and 'p' refer to the retentate and permeate conditions respectively)

Now, with the pervaporation, the retentate side *i.e.*, the feed side of the system are in gaseous state, while the permeate side is in the liquid form. Therefore, the equation 3.13 can be re-written as,

$$T\sigma = -\sum_k J_k^D RT\left(\ln f_k^v - \ln f_k^l\right) - NV\Delta P \quad (3.14)$$

(Here 'v' and 'l' refer to the vapor and liquid conditions respectively)

Also,

$$\left.\begin{aligned} f_k^v &= \varphi_k y_k P = y_k P\left(\text{Assuming } \varphi_k = 1\right) \\ f_k^l &= \gamma_k x_k P_k^S \end{aligned}\right\} \quad (3.15)$$

Combining equations 3.14 and 3.15,

$$T\sigma = -\sum_k J_k^D RT \ln\left(\frac{y_k P}{\gamma_k x_k P_k^S}\right) - NV\Delta P \text{ (Hwang, 2004)} \tag{3.16}$$

Now, in the form of Onsager's dissipating energy relationship, the equation can be written as,

$$T\sigma = -\sum_k N_k RT \ln\left(\frac{y_k P}{\gamma_k x_k P_k^S}\right) = \sum_k \vec{J}_k \vec{X}_k \tag{3.17}$$

Now for a binary system pervaporation process we can say there are two components are exchanging across the membrane. The flux of each component is given by $\vec{J}_k = N_k$ and the force term is given by $\vec{X}_k = -RT \ln\left(\frac{y_k P}{\gamma_k x_k P_k^S}\right)$. Now rearranging the force term , we can say,

$$\vec{X}_k = -RT \ln\left(\frac{y_k P}{\gamma_k x_k P_k^S}\right) \tag{3.18}$$

$$\vec{X}_k = -RT \frac{\left(y_k P - \gamma_k x_k P_k^S\right)}{\mathrm{lm}\left(\Delta f_k\right)} \tag{3.19}$$

Now, with the Onsager's linear relationships and neglecting the off-diagonal phenomenological constants (as we know the primary driving force here in pervaporation is the concentration differences evolved pertaining to the VLE). Therefore,

$$N_k = L_{kk} \vec{X}_k = -L_{kk} RT \frac{\left(y_k P - \gamma_k x_k P_k^S\right)}{\mathrm{lm}\left(\Delta f_k\right)} = Q_k \left(y_k P - \gamma_k x_k P_k^S\right) \tag{3.20}$$

Equation 3.20 shows the basic equation for pervaporation and Qk is called the permeability.

Symbols used in the chapter

P_A : Partial pressure of component A in vapour phase (N/m^2)

P_B : Partial pressure of component A in vapour phase (N/m^2)

x_A : Mole fraction of component A in liquid phase

x_B : Mole fraction of component B in liquid phase

y_A : Mole fraction of component A in vapour phase

y_B : Mole fraction of component B in vapour phase

L_{ij} : Phenomenological Coefficient

X : Force which initiates flux

σ : Rate of change of entropy (J/K.s)

Δc : Change in the molar concentration (mol/m^3)

ΔT : Change in temperature of the system (K)

N_k : Number of moles of component 'k' (moles)

$\Delta\dot{\mu}_k$: Rate of molar free energy change (J/mol.s)

J_k^0 : Diffusional flux ($mol/m^2.s$)

P : Total pressure of the system (N/m^2)

3.3. CHAPTER IN BRIEF

- Concept of bioprocess engineering.
- Production process for biobutanol and ABE fermentation.
- Pervaporation - concept and application.
- Basics of non-equilibrium thermodynamics and its application in pervaporation.

Chapter 4

Application of Membrane in Biomedical Processes

So far what we have discussed is the application of membrane in different industrial sectors, where the intricacy lies in selecting the membrane process as one of the efficient technologies that gives best results. However, in recent days one of the finest usages of the membrane separation in life saving applications like dialysis, artificial breathing support or in brief artificial life support system. The critical part of designing membrane separation process in these cases is the consideration of the sterility of the process and the expenses apart from the best outcome, in order to reduce the total medication costs. Since, in the era of 50's the chemical engineers were being closely associated with the biomedical communities to device scaffolds for artificial organs along with several biochemical reactions going on inside our body. One of the promising breakthroughs in this field is formulation of biocompatible materials that are used to fabricate artificial organs such as valves, prosthetic devices, arteries, veins etc. leading to an enhanced lifetime. In this chapter, some of these biomedical applications of membrane will be discussed in order to initiate the keynote thought among the communities on further development in this field.

4.1. ROLE OF A CHEMICAL ENGINEER CONTRIBUTING TO BIOMEDICAL APPLICATIONS

Before going into the detailed discussion on the application of membrane in biomedical field, the key question comes in our mind is what is biomedical engineering? How this is different from chemical engineering? Whether the two forms can be amalgamated or not in contributing towards the society?

The answer to the first question is itself the name of the field – it is an engineering that connects the acquisition of engineering science to biomedical science improving the human life after clinical trials. It helps to acquire the knowledge of a living system after experimental process and thereafter, substantiates the effort for its betterment through medications. On the contrary chemical engineering is a subject that helps to device anything, which helps to substantiate the betterment effort through knowledge on the materials' property, biochemical reactions and the related mathematics. Let us take an example on such cross-disciplinary areas in order to understand the correlation between two aforementioned engineering subjects and their individual role. In recent days the most critical part in saving a patient's life during hospitalisation is the control of pathogenic hospital acquired (called nosocomial) infection that becomes life threatening. The pathogens are mutant[1] and antibiotic resistant that the application of multidrug fails to act. The role play of a biomedical engineer is to understand the nature of pathogen, the antibiotic class that can reduce the infection level and the way drug acts within the living system. On the contrary, the chemical engineer will understand the formulating material for the drug, the mechanistic part during the course of the drug action within the living system. However, the common part for both the engineers will be the reaction sequences that are happening during drug action that develop a bridge between a chemical engineering and the biomedical

[1] *A new organism generating after being mutation in nucleotide sequences following the original trait along with some additional features.*

engineering as shown the Fig. 4.1. Looking at the history on the involvement of chemical engineering science in biomedical science properly reflects the long term association between these two disciplines for the benefits of the society.

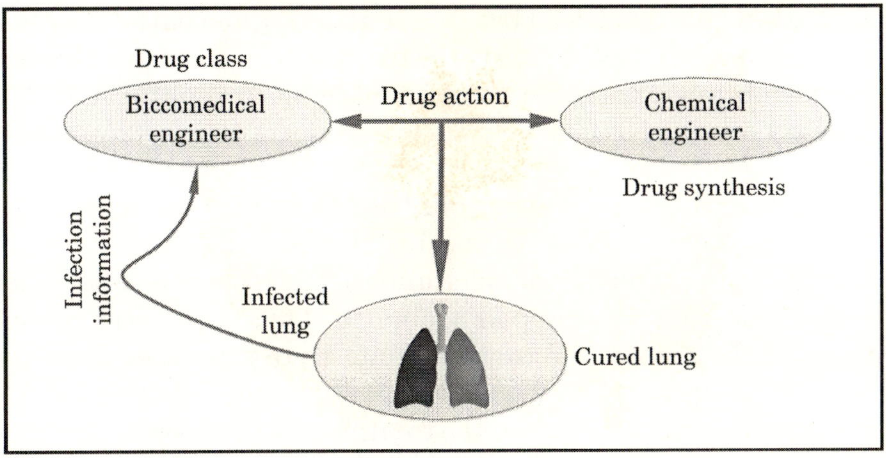

Fig. 4.1: Role of Chemical and Biomedical engineer

Some of the applications are explained in the following sections along with their merits and demerits of the presently used membranes, their costing and questions on the sterility of these membranes in using as biomedical tools.

4.2. DIALYSIS MEMBRANES

Thomas Graham, Professor of Chemistry at Anderson's University in Glasgow, first used the term dialysis in the year of 1861. According to him, the first dialysis process he observed is the diffusion of the crystalloids through vegetable parchment coated with albumin, which behaved as a semi-permeable membrane. With this conception he was trying to extract urea from urine[2]. In 1913, Abel, Rowntree, and Turner from John Hopkins Medical

[2] *Source:* http://renux.dmed.ed.ac.uk/EdREN/Unitbits/historyweb/HDWorld.html

College fabricated the first artificial kidney, where their membranes are made up of treated parchment (Paul, 2006). In 1943, WJ Kolff ("Father of Dialysis") and H Berk developed the first practical human haemodialysis machine. In their fabrication they had used a cellophane tubing of 45 cm with 25 ml of blood fixing it on a small board followed by rocking it in a saline bath. They had added 400 mg/100 ml of urea in the blood which had passed through the membrane after ½ an hour from the blood (Kolff, 1965). The first kidney dialysis clinical trial was performed by a 42 years old Swedish professor Dr. Nils Alwall at the Department of Internal Medicine at the University hospital in Lund[3]. The dialysis machine he had made, was comprised of cylindrical glass container contains a large wire mesh made up of stainless steel. On this a flattened cellophane tube is wrapped around it. Once the blood passed through the casing, the contaminants were separated from the blood. The first chronic dialysis patient went for dialysis clinical trials on March 9, 1960 under the treatment of Dr. Belding Scribner, Dr. Wayne Quinton, and Dr. David Dillard. They had used a Teflon R tubing shunt (Scribner, 2002) (Fig. 4.2), which was supposed to be non-reactive with the tissue. They had performed the dialysis process for 11 hours. In 1961, Scribner met with a Norwegian urologist Fred Kiil, who had fabricated a less resistant flat plate dialyser equipped with cellulosic membrane Cuprophane (Blagg, 2015). In 1962, Cimino and Brescia had developed a veno-venous access for dialysis that used a sphygmomanometer to dilate an accessible forearm vein along with a blood pump. The cleaned blood was returned through another vein, usually in the ankle. The discovery, called "arteriovenous fistula" was found to be phenomenal (Cimino *et al.*, 1962). However, several strategic developments have been done so far in the machinery fabrication for haemodialysis so far. In recent days, scientists are working on designing an artificial transplantable kidney that eliminates the laborious and painful process of dialysis. Dr. Shuvo Roy, Bioengineer from the laboratory at Mission Bay campus of the

[3] *Source:* http://www.med.lu.se/english/about_the_faculty/history_of_the_faculty/personalities_discoveries_and_innovations/nils_alwall

University of California-San Francisco is in a process of fabricating an artificial transplantable kidney of size of a coffee mug along with its collaborators from University of Michigan, the Cleveland Clinic and Case Western Reserve University[4].

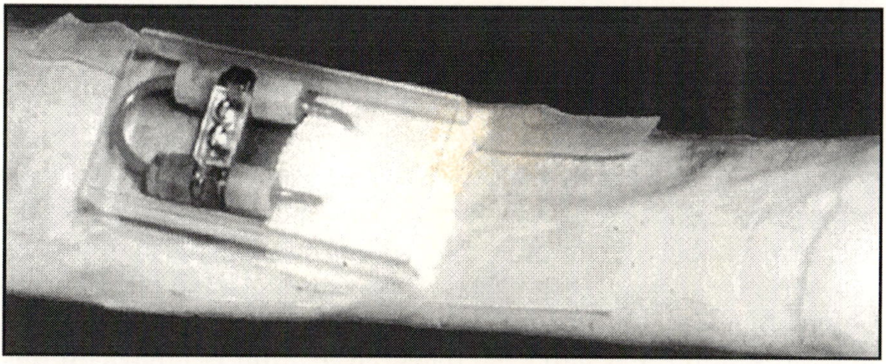

Fig. 4.2: Clyde's U-shaped Teflon shunt after four weeks in place (*Reprinted after the permission taken from Nature Medicine, 8(10), 2002*).

Now before going into the in-depth dialysis process, let us be educated with the happenings in dialysis. Dialysis is the mechanism in case of kidney failures. Now, what kidney does? It has primarily three functions – fluid balance, electrolyte balance and acid/base balance. Therefore, in dialysis, one will do all the performances using a countercurrent flow between dialysate and blood through a hollow fibre channel (as shown in Fig. 4.3). One of the crucial points in dialysis operation is the dialysis membrane that decides the performance of the process in separating the toxic substances from the blood. Moreover, as the blood comes in contact with the membrane, it needs to ensure that there should not be any reaction between the blood immune cell and the membrane material. Further, the membrane flux is also an important factor that controls the permeation of a wide range of solutes having different molecular weight (Eknoyan *et al.*, 2002). According to the author, membranes with high porosity (primarily observed with the synthetic

[4] *Source*: http://science.kqed.org/quest/2010/09/21/scientists-work-to-develop-artificial-kidney/

membrane compared to previously used cellulose membrane) create a passage for β_2-microglobulin[5] (molecular weight 11,900 Dalton and excreted from kidney). On the contrary, low porosity membrane increases the possibility for the increase in the concentration of β_2-microglobulin resulting in an increased mortality among the patients (Mumtaz *et al.*, 2010). However, the complexity arises from the question on the separation of other uremic toxins, such as advanced oxidation protein products (AOPP) and protein carbonyls (PCO). It was seen by researchers (El-Wakil *et al.*, 2011) that even with the high flux membrane the extent of separation for AOPP is unaffected. This indicates a comparative increase in the reduction of mortality caused with the low flux membrane, where all the toxins are accumulating during chronic haemodialysis. In next few sections a thorough discussion on the dialysis membranes will be elaborated in order to have a glimpse on how the chemistry along with physics are involved between the membrane and blood.

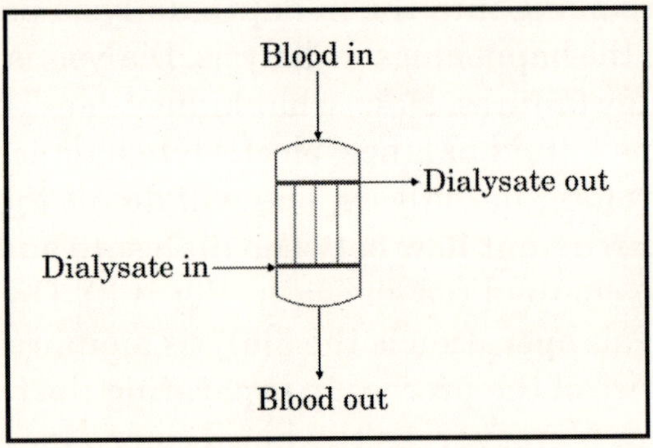

Fig. 4.3: Hollow fibre membrane set-up for dialysis operation

The primary aim for a dialysis membrane is to separate the uremic solutes that are accumulated in the blood due to renal failures and the restoration of depleted compounds like bicarbonate (Bouré and Vanholder, 2004). However, the critical concern in selecting the membrane for dialysis is the consideration of the

[5] *Source*: https://en.wikipedia.org/wiki/Beta-2_microglobulin

biocompatibility of the membrane so that it will not promote the inflammation and high water throughput from the membrane. Researchers (Dinarello, 1992) have shown that the membrane material has a considerable influence on the production of cytokines and hence, the inflammation cascade. Cuprophane membrane, which is made up of cellulose, has higher affinity for the lipid A[6] (Metzger *et al.*, 2012) structure in lipopolysaccharides (LPS) (Alexander and Rietschel, 2001). The major issue with the haemodialysis is the limited sterility of the dialysate that perhaps the reason for the contamination of blood. In general the membranes that are used for dialysis are primarily AN69 (copolymer of acrylonitrile and sodium methallylsulfonate), cuprophan (cellulosic) and polysulfone (Thomas *et al.*, 2011). One of the common properties with all such membranes exhibits hydrophilic in nature so that the membrane should have some hydrogel structure attributing to the permeation property of the membrane. Hydrogel is basically defined as a three dimensional network structure that can absorb and hold water. With the hydrophilic property, the surface becomes more enriched with water because of the attachment of water with both hydrophilic and hydrophobic tail and this will reduce the interfacial free energy. Therefore, the surface has a less tendency to adhere protein and cells during dialysis. The presence of sulfonate group (Fig. 4.4) in case of AN69 membrane attracts water that eventually forms a hydrogel structure over the membrane manifesting the membrane more biocompatible and high flux.

Especially, with the AN69 membrane maximum amount of β_2-microglobulin can be eliminated because of the adsorption of the protein over highly sulfonated protein surface, although the serum concentration of β_2-microglobulin cannot be reduced (Thomas *et al.*, 2011). However, it is seen that compared to AN69 and polysulfone membrane the adsorption of LPS is less in cuprophane membrane (Laude-Sharp *et al.*, 1990). According to the study, they

---

[6] *Source:* http://www.nature.com/nsmb/journal/v19/n11/fig_tab/nsmb.2393_F1.html

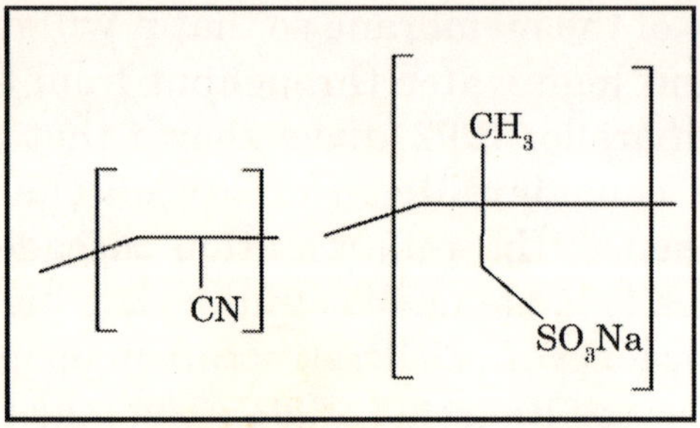

Fig. 4.4: Lively group on the surface of AN69 membrane

had seen that LPS eluted in the membrane washing solvent was mere 2% with cuprophane membrane, marginally comparable with the AN69 membrane (3.5%) and much higher with 9% compared to polysulfone membrane. This says that AN69 is more applicable compared to others as it has the high flux and less chances of infection due to less adhesion of LPS over the membrane. Fig. 4.5 shows a schematic for dialysis and associated membranes.

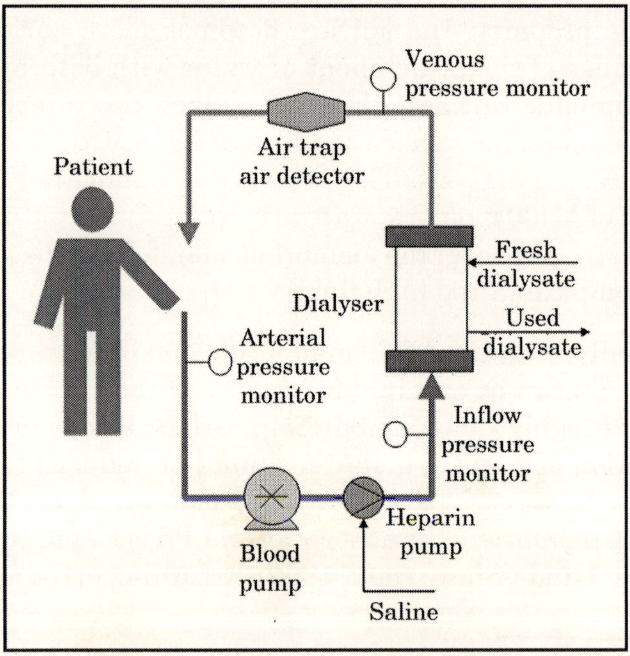

Fig. 4.5: Dialysis operation schematic

4.3. EXTRACORPOREAL MEMBRANE OXYGENATION (ECMO)

One of the critical medical devices that use membrane as a selective barrier for the removal of toxic materials from blood apart from dialysis is ECMO. Here the blood is continuously in contact with an oxygenator membrane for efficient of oxygen (O_2) in blood and perfusion of carbon-di-oxide (CO_2) in alveoli. The first adult respiratory failure patient treated with ECMO was reported in the decade of 70's. However, the number of respiratory ECMO patients remained small for more than 30 years afterward. Although, the number of ECMO medicated patients are being continuously increased. Moreover, the acceptance of its application in adulthood is increased, while in case of neonatal respiratory failures it is a well acknowledged technique. During the gas exchange between alveolus and blood the reactions are listed in Fig. 4.6.

In case of failure of the respiratory system, like acute respiratory distress syndrome (ARDS), where the lung cannot supply enough oxygen to the blood and this will cause the failures of the organs following septic shock. In this case, basic medication will be to ensure the exchange of oxygen externally and to ensure the adequate pumping of oxygen to the blood. ECMO is one of the facilitated mechanisms through which one can pump oxygen into the blood and strip out CO_2 from it. Fig. 4.7 shows a schematic diagram of ECMO. In 1974, Dr. Robert Bartlett of Orange County Medical Center, USA had first applied ECMO to survive a newborn baby (called "Esperanza"), whose arterial pressure dripped to 12 mm of Hg and had no chances of survival. After three days the baby was completely recovered and after that with several subsequent trials on neonatal by Dr. Bartlett, it was seen that 75% was survived although 90% mortality was predicted[7]. However, at that time ECMO was not tested with the adults and people were

[7] *Source*: http://www.in-isrc.org/2012%20Fall%20Seminar/121030.ISRC.Lectures/ISRC%202012%20-%20Adult%20ECMO.pdf

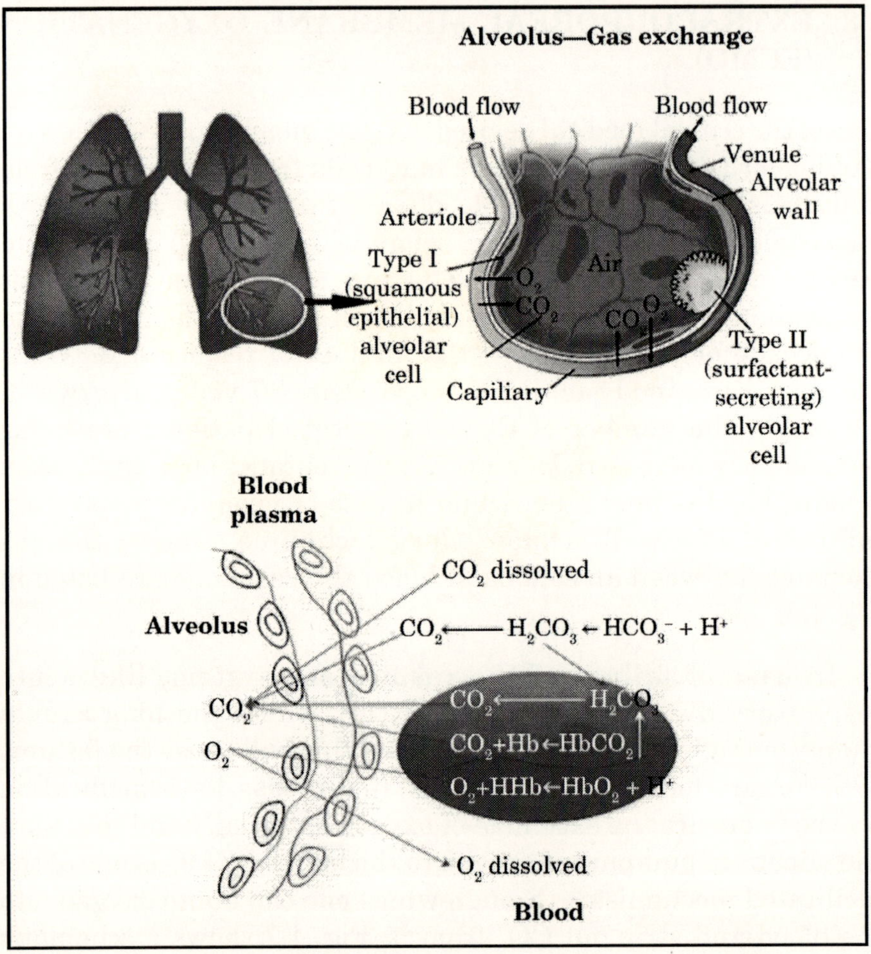

Fig. 4.6: Gas exchange between alveolus and blood

not sure with the adults' respiratory response to ECMO. In 1980, University of Michigan opens their first ECMO center. In the year of 1989, in Boston's Children Hospital, USA around 90% babies were survived with the ECMO, while the 67% were survived with the conventional therapy. In 1996, in United Kingdom, the survival was 70% with ECMO compared to 41% with conventional.

In 2009, in United Kingdom, Conventional Ventilation or ECMO for Severe Adult Respiratory Failure (CESAR) was formulated a

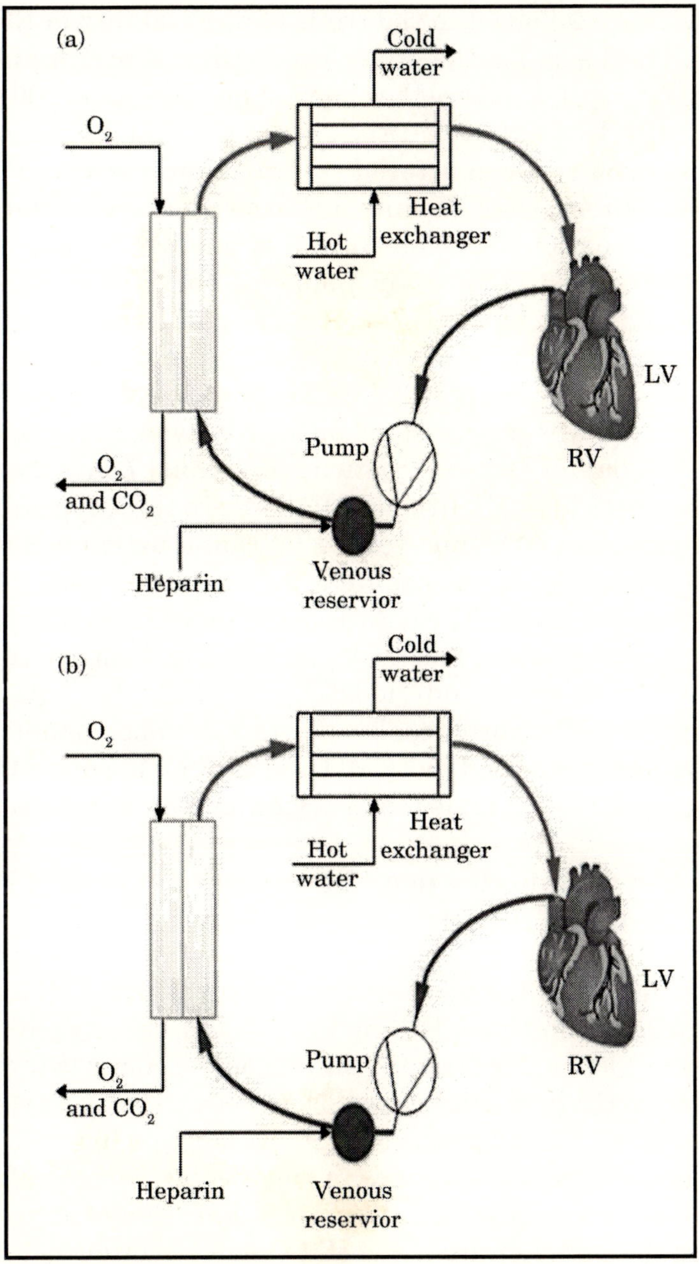

Fig. 4.7: (a) Veno-Arterial (VA) ECMO (b) Veno-Venous (VV) ECMO

team who are involved in the trials of applicability of ECMO for adults. The basis for the study was with a report by Glenfield Hospital group, UK, where they had applied ECMO on 200 patients since 1989. It was seen that the 34% of the first 50 patients died with the ratio between arterial O_2 partial pressure and inspired O_2 partial pressure of 65 mm Hg[8]. However, their study shows death of the patient or severe disability among the patients at six months or before the discharge from hospital (Peek *et al.*, 2006). It was revived again in 2009, when Centers for Disease Control and Prevention (CDC) raised a high alert on H1N1 influenza virus. According to Extracorporeal Life Support Organization (ELSO), 72% survival was observed, when ECMO was instituted with 6 days of intubation. If the intubation goes beyond 7 days the survival rate decreases to 31%[9]. In general, there are three types of ECMO – veno-arterial (VA), veno-venous (VV) and arterio-venous (AV) ECMO depending on the cannulation techniques used (Martinez and Vuylsteke, 2012). The details of the procedures regarding the ECMO techniques and how they are different from each other are described in Martinez and Vuylsteke (2012). In the context of this chapter, we will be purely confined with a thorough understanding of the membranes used in ECMO and the complication that may happen with ECMO under the purview of membrane used here.

4.3.1. Terminologies used in Oxygen Transport Process through Blood

Fig. 4.8 describes the mechanism of oxygen-carbon-di-oxide (O_2/CO_2) exchange in blood. During the process, the oxygen pressure reduces to 100 mm Hg from 150 mm Hg once it comes to the alveolar region from the bronchial region. Now within the blood vessel, the deoxygenated blood has the oxygen pressure of 40 mm Hg, lower than the oxygen pressure within the alvolar portion. Therefore, the vessels encapping the alveoli, the diffusion of oxygen takes place in the direction from alveoli to vessel in order to enrich the

[8] *Source*: http://cesar.lshtm.ac.uk/summaryprotocolrevisedfeb2003.pdf

[9] *Source*: https://www.elso.org/Registry/H1N1Registry.aspx

oxygen within the blood. Now, how the diffusion mechanism takes place within the alveolus and what are the issues associated with the transfer? Once the oxygen is transported to the blood, four oxygen molecules bind with haemoglobin (100% saturated). Similarly if three oxygen molecules will bind it will be 75% saturated haemoglobin. In an adult male the normal range for haemoglobin will be 14–16 g/100 mL, while for females this is around 12–15 g/100 mL of blood. Each gram of haemoglobin is capable of carrying 1.34–1.39 mL of oxygen. Therefore, the volume percentage of oxygen in blood will be (1.34–1.39)×(14–16) mL of oxygen per mL of blood for 100% saturated haemoglobin.

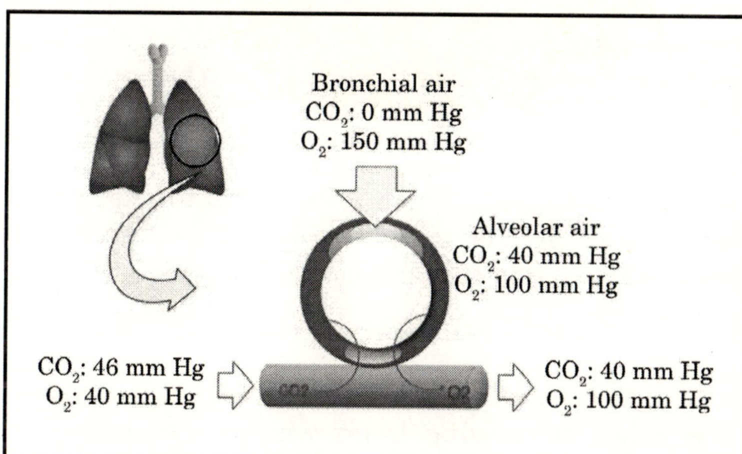

Fig. 4.8: Mechanism of oxygen-carbon-di-oxide (O_2/CO_2) exchange in blood

Now if the haemoglobin saturation will be, say, 97%, the volume percentage of oxygen will be (1.34–1.39)×(14–16)´0.97 mL of oxygen per mL of blood. Now, at normal body temperature the olubility of oxygen within the blood is around 0.003 mL of oxygen in 100 mL of blood for every 1 mm of Hg of partial pressure of oxygen (P_{O_2}). For 100 mm of HgP_{O_2} the solubility will be 0.3 mL of oxygen per 100 mL of blood. In general 2–3% of the oxygen is in the dissolved condition. In case of oxygen transport process, the total arterial, venous and palmonary capillary volume of oxygen per 100 mL of blood are given in equation 4.1a to 4.1c below.

$$Ca_{O_2} = \left[(1.34 \text{ to } 1.39)\times(14 \text{ to } 16)\times Sa_{O_2}\right]+\left(Pa_{O_2}\times 0.003\right) \quad (4.1a)$$

$$Cv_{O_2} = \left[(1.34 \text{ to } 1.39)\times(14 \text{ to } 16)\times Sv_{O_2}\right]+\left(Pv_{O_2}\times 0.003\right) \quad (4.1b)$$

$$Cc_{O_2} = \left[(1.34 \text{ to } 1.39)\times(14 \text{ to } 16)\times Sc_{O_2}\right]+\left(PA_{O_2}\times 0.003\right) \quad (4.1c)$$

The first part of the above expressions are the volume of oxygen binds to haemoglobin and the second part is the amount of disoolved oxygen in blood serum. There are other two terms called delivered oxygen (D_{O_2}) and oxygen consumed by the tissues (V_{O_2}), which is primarily the oxygen uptake by the cell. If the volume of the blood pumped by heart per minute, called cardiac output (CO), is known then the delivered oxygen is given as D_{O_2}=CO×Ca_{O_2}. In general the formula for the delivered oxygen is given by equation 4.2.

$$D_{O_2} = CO\times Ca_{O_2} \times 100 \text{ (unit: mL of oxygen per minute)} \quad (4.2)$$

Now if the delivered oxygen volume will be expressed as the volume of oxygen per 100 mL of blood per m^2 of the body surface area, then the equation 4.2 can also be written as,

$$DI_{O_2} = CI\times Ca_{O_2} \times 10^5 \text{ (unit: mL of oxygen per minute per } m^2 \text{ of the body surface area)} \quad (4.3)$$

'CI' is called the cardiac index which is the ratio of cardiac output per m^2 of the body surface area. The normal CI value is 2.5 to 4 L of blood per minute per m^2 of body surface area. The normal CO value is 4900 mL of blood per minute.

One of the basic primary features in oxygen transport is the binding of haemoglobin, which is called oxygen-haemoglobin kinetics given in 4.4.

$$O_2 + Hb \underset{k}{\overset{k}{\longrightarrow}} HbO_2 \quad (4.4)$$

Therefore, the rate of oxygenation can be written as,

$$-\frac{d[O_2]}{dt} = -\frac{d[Hb]}{dt} = \frac{d[HbO_2]}{dt} = k[O_2][Hb] - k'[HbO_2] \quad (4.5)$$

At equillibrium $\frac{d[O_2]}{dt} = \frac{d[Hb]}{dt} = \frac{d[HbO_2]}{dt} = 0$. Therefore, the rate equation can be given by (4.6).

$$\left.\begin{aligned} & k[O_2][Hb] - k'[HbO_2] = 0 \\ & \Rightarrow \frac{[HbO_2]}{[Hb]} = \frac{k[O_2]}{k'} = K[O_2] \end{aligned}\right\} \quad (4.6)$$

The fractional saturation of haemoglobin is given by

$$S = \frac{[HbO_2]}{[Hb]+[HbO_2]} = \frac{K[O_2]}{1+K[O_2]} \quad (4.7)$$

Equation 4.7 is called the equillibrium realationship for the oxygenation of haemoglobin.

4.4. OTHER BIOMEDICAL APPLICATIONS OF MEMBRANE

In present days, one of the most demanding applications of membrane emerges in the field of nanofibrils. The enhanced bioactivity of the membrane increases the interest on its application at different level of biomedical fields. Cellulosic nanomaterials is one of such keys to develop applications dependent solution against different biomedical fields. Before saying why to nanocellulosic structure, let us first have some brief idea on "*nano*". During discussion on nano, the first two questions that are coming up is, "*What* is nano? And *Why* nano?"

Nano is something with the molecular scale. In 1980, Dr. K. Eric Drexler coined the concept of nano-science, where he was saying about the miniatured machines at a scale of nano-metres. According to National Nanotechnology Initiative, USA something which is less than the scale of 100 nano-metres is called nano-materials. Rather the technology is a bottom-up approach, where the people are continuously building up complex things after accumulating the things from the atomic level. Fig. 4.9 gives some concept of this bottom-up approach.

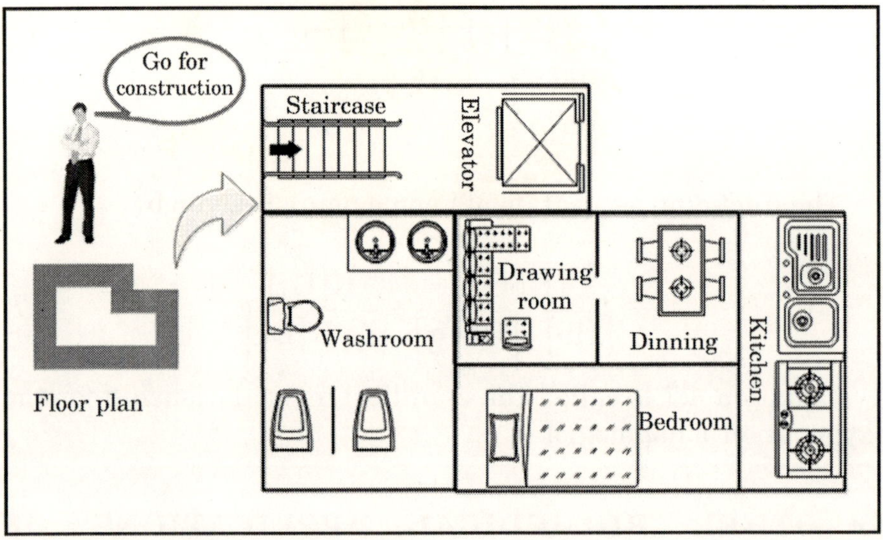

Fig. 4.9: Building block of a process deliverable to end user

So in what aspect nano-materials are active? Nano-materials are those, which are divided into tiny segments and then only one can get the nanoscaled material. Now once, the bulk is reduced to tiny one, the more and more surface area will be exposed. Therefore surface to volume ratio becomes large indicating that the more atoms are at the surface level. Because of the small sizes in a very higher level we can say that the more and more atom will be exposed to the surface that makes the nano a reactive one. At the same time, the reactivity mainly arises due to the size of the material. If it is too small, the electrons are confined, band gap energy increases and hence, the shifting of the electron from ground state to conduction band will be ceased. Therefore, once we will go for the active nano sheet we have to ensure the size of the matter also. Now, the question is how the membrane could be fitted to the nano-structure level and how this could be applied to biomedical?

One of the typical membranes, called nano-composite membrane, is a footstep area in the membrane field. Now what is nano-composite? Nano-composites are those amalgammed solid components, where at least one component is at nano-scale and has the repeating distanced between the other components. In true sense, instead of saying it amalgammed matter, we can call it a multiphase solid material. Fig. 4.10 shows some schematic view of the nano-composite.

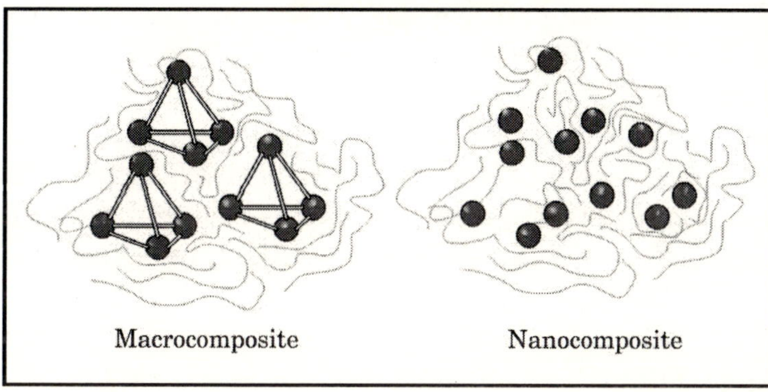

Fig. 4.10: Macro and Nano-composite matrix

Nano-composite membrane has numerous applications. One of the most signifcant applications found is in the imaging and drug delivery system. In drug delivery applications the core sheel technology is the most sophiosticated one that is widely applied for drug sysnthesis. Fig. 4.11 shows such technology. Recently in a publication by Furtos *et al.* (2016), antibiotic amoxicillin is encapped within the membrane to understand its release from the coreshell in mitigating the detal issues.

The core shell technology attched with the imaging system is the encapsulation of the metals or its alloys within a biocompatible

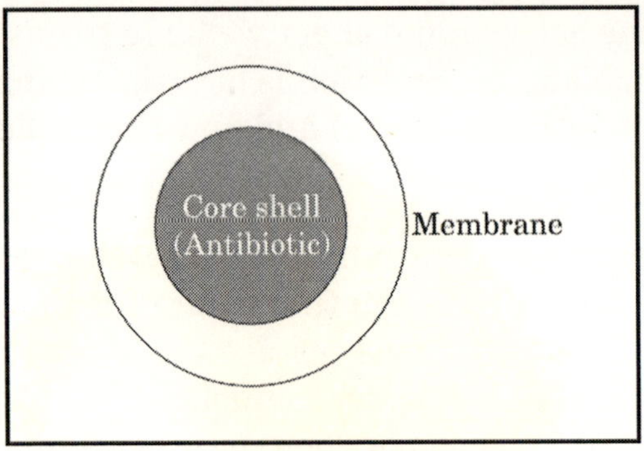

Fig. 4.11: Core-shell technology for encapsulation of drug within nanocomposite membrane

structure on which a bioactive satellite structure is embedded. Now, in case of bioimaging, especially with the magnetic resonance imaging (MRI), the H^+ ions present in the water tissue is alligned in the direction of applied magnetic field. Now applying a strong resonance field perpendicular to the magnetic field, yield a net directional allignment of H^+, which upon release relaxed the ions in the magnetic field direction. Now, this relaxation is called T_1 relaxation and it is different for different fluids from different tissues. With the contrast used, this relaxation will be varied and hence the image gets cleared with proper visibility. In a recent study (Qazi *et al.*, 2013), multiple chelated Gd^{3+} (Gadolinium) ions are attached within the interior of bacteriophage P22 viral, which enables nanoscale MRI contrast agents having high relaxivity values (Fig. 4.12). One of the typical morphological utilities that we have with the P22 is its highly fenestrated "wiffleball" texture, acts like a membrane, ensuring efficient water (H^+) exchange between the environment and interior cavity of the capsid.

4.5. CHAPTER IN BRIEF

- Application of membrane in biomedical process.
- Dialysis process basics.
- Dialysis membrane basics.

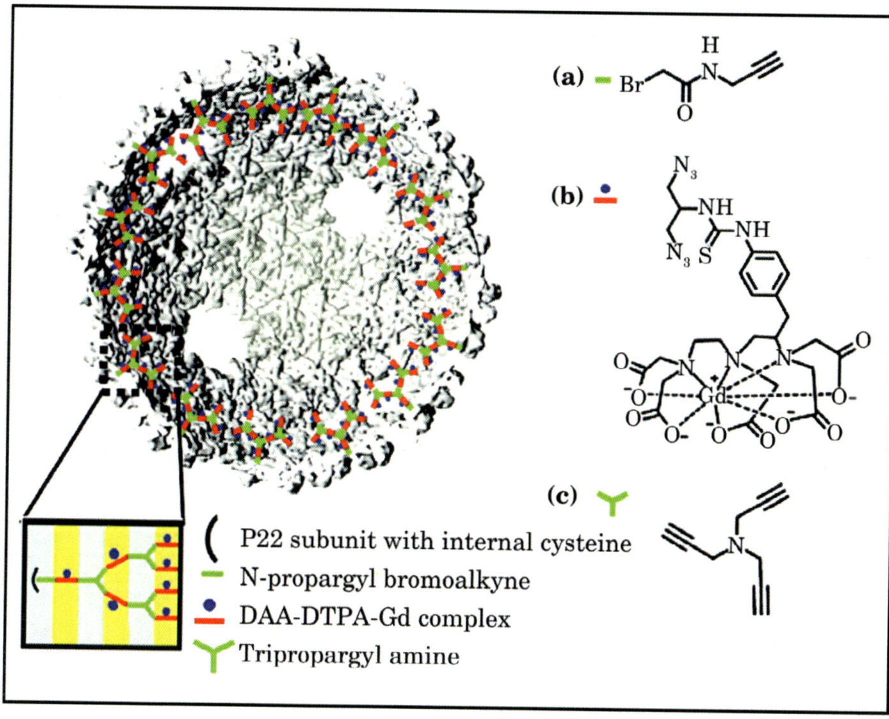

Fig. 4.12: Theoretical incorporation of seven DAA"DTPA"Gd complexes per subunit of P22 WB. Monomers used for synthesis of branched oligomer (a) N-propargyl bromoalkyne, (b) 2-azido-1-azidomethyl-ethylamine coupled to 2-(4-isothiocyanatobenzyl)-diethylene-triamine pentaacetic acid gadolinium (DAA"DTPA"Gd complex), (c) tripropargyl amine (*Reprinted with permission from Qazi, S., Liepold, L.O., Abedin, M.J., Johnson, B., Prevelige, P., Frank, J.A., Douglas, T., 2013. P22 Viral Capsids As Nanocomposite High-Relaxivity Mri Contrast Agents. Mol. Pharm. 10, 11–17. Copyright 2012 American Chemical Society*)

- ECMO process and the membrane application in ECMO.
- Oxygen transport mechanism in lung.
- Nano-composite membrane and its application in bioimaging techniques as contrast.

Chapter 5

Application of Membrane as Membrane Reactor

So far we have discussed different applications of membrane in different sectors of engineering, sometimes with some specialized applications. Membrane reactor is something that can be accommodated in any of the applications described in previous chapters. The first application of membrane reactor comes in conjunction with the activated sludge process. The first coupling of membrane and activated sludge process was researched by Rensselaer Polytechnic Institute, Troy, New York and Dorr-Olivier, Inc. Milford, Connecticut, US in late 60's. What they have done? In their system, they had fed the liquor comprising of biomass and waste water to a membrane, where the concentrate was again recycled back to the reactor or aeration tank and permeate was passed through the membrane. However, the major disadvantages comes with the process is its high cost and low longevity of the membrane because of fouling. The people look into good opportunity to design some process, where the fouling could be reduced so that the membrane can be reused many times. This reduces the cost of the membrane procurement. The breakthrough comes in 1989, when the membrane was submerged into the reactor just above the air diffuser. This arrangement provides a scouring over the membrane because of continuous aeration and thus formulates a washing process for deposition over the membrane surface. Now, the foremost question is what is membrane reactor? How it becomes superior to conventional membrane separation process.

According to International Union of Pure and Applied Chemistry (IUPAC) "*The membrane reactor is a device that combines the separation properties of membrane with the typical characteristics of catalytic reaction steps in only one unit*" (Mallada and Menéndez, 2008). Fig 5.1 shows different application based membrane reactor set-up.

In the figure apart from 5.1(a), all the membrane reactors (shown in Fig. 5.1(b), (c) and (d)) are designed as a sole module, where the both the intended action and the separation are served simultaneously. The reader can understand because of such utility the whole process cost can be minimized as the low number of installations reduces the piping cost, land cost, manpower cost, housing cost, maintenance cost etc. Now in this chapter, we will try to understand the membrane reactor based on its application at different sectors. Primarily, we will keep ourselves restricted to its applications – (i) Wastewater treatment (ii) Food industry and (iii) Alternative fuel industry.

5.1. APPLICATIONS OF MEMBRANE REACTOR IN WASTEWATER TREATMENT

In ancient time the civilization that probably first felt the need for wastewater treatment was Rome, where the population had attempted to reduce the pollutants load in water. In 1500 BC, Mohenjo-daro civilization equipped with the concept collection of waste water and rain water separately. Now the collection of wastewater and pouring them into the river is not a solution. Therefore, we need a gross understanding how the pollutant load can be reduced in the wastewater before it is discharged. This key understanding of wastewater treatment paved a pathway to

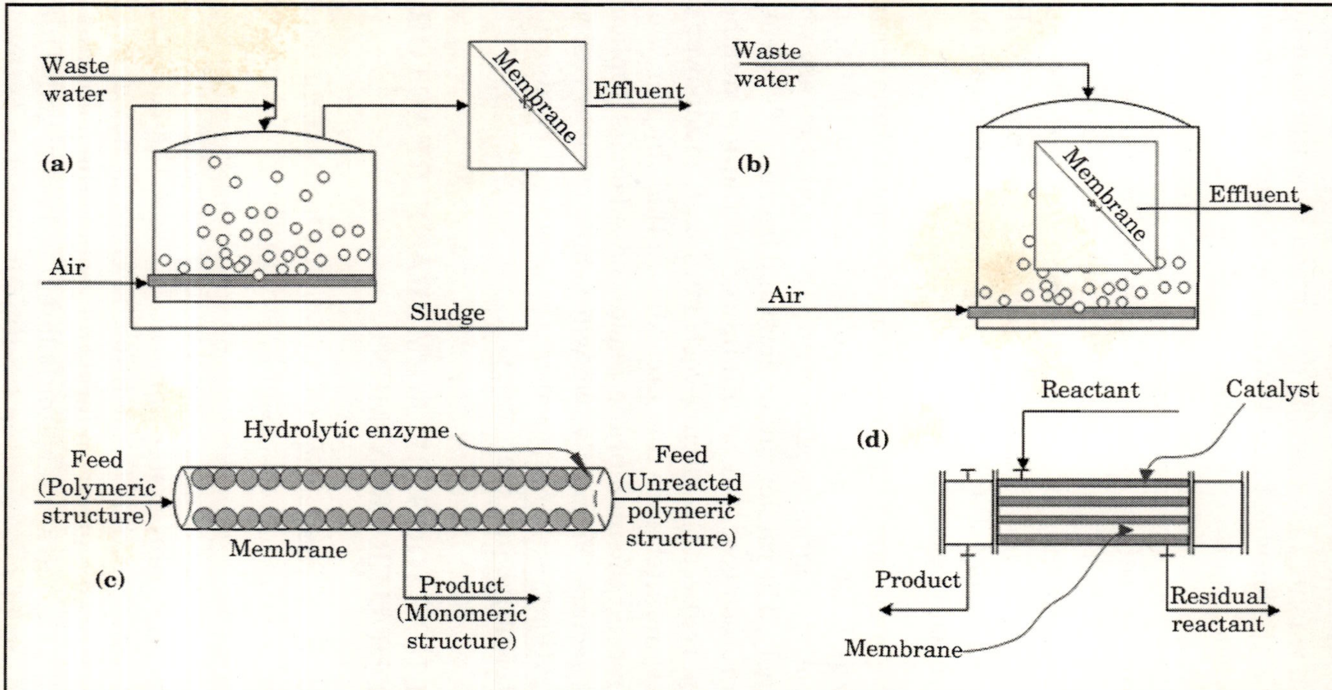

Fig. 5.1: Different membrane reactor set-up (a) Side stream membrane reactor for activated sludge process (b) Submerged membrane reactor for activated sludge process (c) Immobilized membrane reactor (d) Catalytic membrane reactor

develop new technologies that are coming up now. The first concept of treatment was developed in 19th century, when the wastewater was flown down to irrigation channels and once, water seeped through the ground, which is collected to discharge it into the canals. Now if you look at the process of clarifying water through seepage is what we call filtration. In modern days the process becomes sophisticated with the membrane separation process. The most conventional wastewater treatment process is activated sludge process, where the wastewater is fed into an aerated reactor comprising of sludge having different bacterial contaminations in order to remove the organics from wastewater. Upon clarification using aerated reactor the stream is fed to the clarifier in order to settle down the sludge and remove the clarified water as the top effluent from the clarifier (Von Sperling, 2007). Fig. 5.2 shows such activated sludge process.

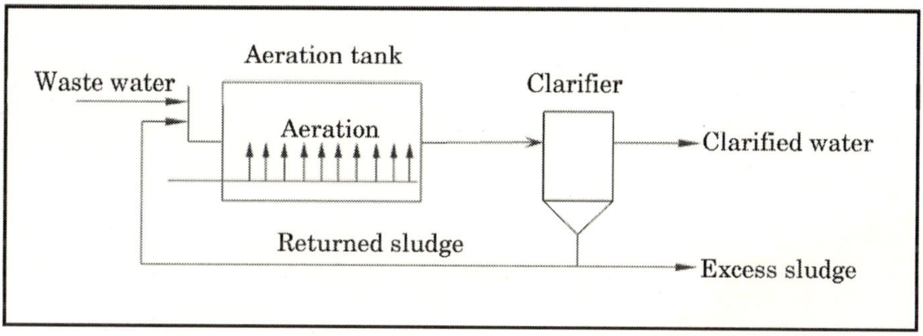

Fig. 5.2: Schematic of activated sludge process

Now, the question comes, where we shall include the membrane for the same. As we discussed in our previous chapter, the membrane is something to separate two miscible or immiscible solids from the liquid stream. However, the operability of a clarifier is limited to a large volume handling. The first commercial side stream membrane bioreactor comes in 70's introduced by Dorr-Oliver (Jenkins and Wanner, 2014). The side stream of bioreactor (shown in Fig. 5.1a) goes into a membrane, where the sludge is separated out from the product stream (effluent) to recycle it to the reactor. One of the obvious utilities of side stream bioreactor is the less interaction time of membrane with the wastewater components and has a less possibility of sticking the material on the surface of the membrane. Now the question comes – "*How far the statement is correct one?*". As said before, the fouling of the membrane is the crucial point of discussion that controls the membrane process performance. Turbulence over the membrane might be advantageous, which mechanically cleans the membrane and maintain the membrane throughput. In case of the side stream bioreactor the turbulence promoter is not so high enough, which can actually reduce the deposition over the surface and subsequently reduce the fouling. The increase in the cross-flow velocity might serve the purpose to some extent. However, if the velocity is very high it will have some abrasion on the surface and may result the puncture of the membrane. Therefore, some sort of arrangement should be made so that there must be substantial turbulence over the surface without being much detrimental to the membrane. Submerged membrane bioreactor acts as turbulence promoter in order to have substantially low fouling of the membrane. Fig. 5.3 shows a schematic view of the submerged membrane reactor, sometimes which is also called "*submerged aerated membrane bioreactor*".

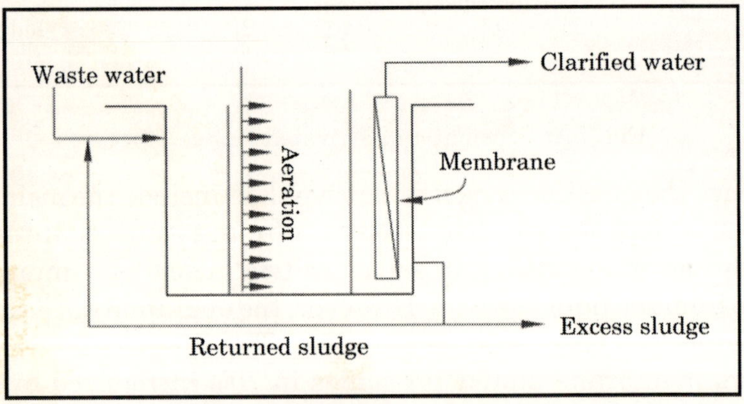

Fig. 5.3: Submerged membrane reactor

In Fig. 5.3 the membrane is confined to a different chamber, where the aerated water goes into the chamber and the sludge is being separated through membrane as the retentate. The clarified water comes out as permeate. As clearly understood, with the arrangement, the membrane fouling will be a major issue in this case. Here, we have no turbulence over the membrane, which could eliminate the deposition problem and thus the fouling. Fig. 5.4 shows an arrangement for submerged membrane reactor, where the membrane and the aeration will be done within the same chamber. Therefore, enhanced turbulence over the membrane surface because of constant aeration will help to eliminate the fouling problem.

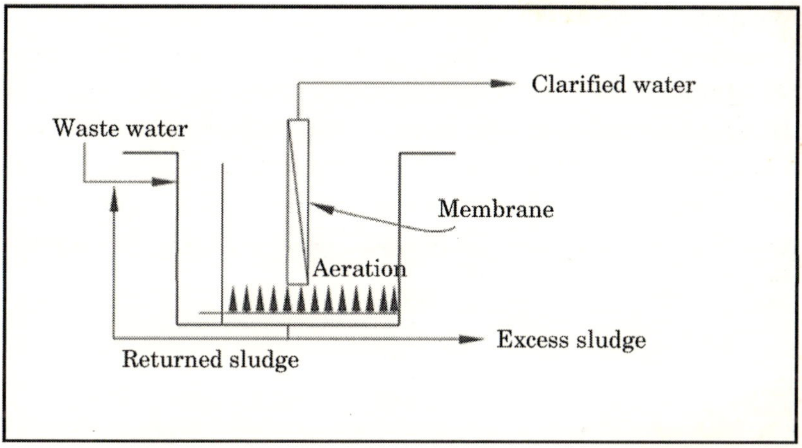

Fig. 5.4: Submerged membrane reactor where membrane is in the reaction chamber

5.2. BASIC CONSIDERATIONS IN SUBMERGED MEMBRANE REACTOR OPERATION AND DESIGN

The primary consideration in the fabrication of such a reactor for the treatment of waste water can be estimated based on the following estimations.

- Preparation of reactor condition and thereafter its maintenance.
- Flow rate of aeration.

- Maximum and average flux throughput from the membrane.
- Mixed liquor suspended solids (MLSS) in the aeration tank.

In the preparation of activation sludge tank, the primary consideration is the biochemical oxygen demand (BOD) of the effluent coming to the tank as the primary effluent. This is what we are calling as the food to the microorganisms present in the tank. Like human, in order to consume food we need oxygen; the microorganisms also demand oxygen, which is supplied through aeration. Therefore, to have a control over the process and efficient reduction of BOD, there should be a ratio between foods (F) to microorganism (M) maintained in the system. Microorganisms are such amount, which is basically the volatile matter measured in MLSS. Hence, if in an area, the average sewerage flow is F_{av} GDP and BOD_5 is X mg/L, the food amount will be $3.8F_{av}X$ mg/d (1 gallon = 3.8 Litre). Again if the volatile matter will be VM mg/L and the volume of the effluent to be processed is V, the microorganism's amount will be said as (V) (VM) mg. Therefore, the ratio per day will be given as,

$$\frac{F}{M} = \frac{3.8F_{av}X}{(V)(VM)} \text{ per day} \tag{5.1}$$

For conventional sludge process, this ratio is in the range 0.25 to 0.45, while with the extended aeration activated sludge this ratio becomes 0.05 to 0.15. In order to understand the ratio change let us start with the Fig. 5.5. In the endogenous growth region the stored food metabolises and organisms feed one another. This is because of the depleted food to the system as we need a specific time to reduce the BOD of the system. Therefore, from the figure it is very much evident that the sludge production decreases at this particular zone as the consumption of food is limited. Therefore, at this stage we need more aeration with enormous cell growth in order to mitigate the incoming BOD. Hence, the ratio will be decreased at this stage.

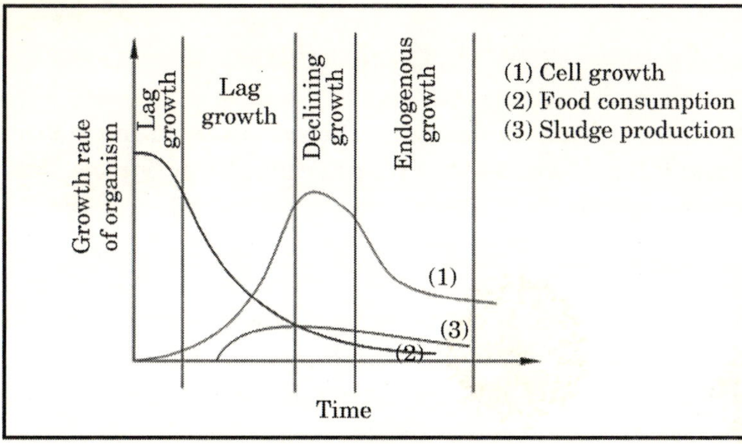

Fig. 5.5: Food (F) depletion, Microorganisms (M) growth rate and sludge production variation with time

Hence, one of the important aspects of the process is the cell retention time (CRT) within the tank. Rather, we may say Fig. 5.5 shows that the activities of microorganisms in a biological treatment system attributes to the average age of the organisms in the system or the "CRT" of the system.

$$\text{CRT} = \frac{(\text{V})(\text{VM})}{\text{F}_{\text{sludge}}} \text{ day} \tag{5.2}$$

Where, F_{sludge} is the excess sludge production (kg per day).

For conventional one the value of CRT is 4-6 days, while for extended aeration activated sludge 15-25 days. It is obvious that at the end, with the declined condition of microorganisms' growth, the value for retention time increases in order to maintain the effluent treatment process spontaneous. Now the main idea with the CRT is that we need to increase CRT in order to make the treatment process rigorous. Now, with the membrane bioreactor set-up, the CRT will be increased as the retention of the biomass will be there in the system. While, if the system will be the conventional one, there is a parameter, which needs to be optimized. This is called hydraulic retention time (HRT).

$$HRT = \frac{V}{3.8F_{av}} \text{ day} \tag{5.3}$$

Now let us begin to understand the system with the modulation of CRT and HRT. If we say that we have small CRT or the retention of the microorganisms is less, it means we have increased flow for excess sludge. Now for a process with clarifier, if we want to increase CRT, that means we need to increase the volume of the effluent, which is nothing but the food. If we do so that means the HRT we also get increased. This can be compensated, once we have the SMBR set-up. This will increase CRT without the increase in HRT. Hence, the main controllable parameter here is that excess sludge rate. Or in other words, CRT for the system is controlled/ maintained by wasting the appropriate amount of excess biomass–called waste activated sludge (WAS). Let us go with a problem.

Example 5.1[1]

A cell residence time of 5.8 days is desired. With 5800 pounds of microorganisms in the aeration tanks, calculate the pounds of organisms that must be wasted per day.

Applying eq. 5.2, organisms wasted per day = $\frac{5800}{5.8}$ = 1000 lb per day

Another important parameter to judge the process efficiency is the yield coefficient of the process. The more and more biological growth in the system, it will ensure proper reduction of BOD from waste stream after primary unit. Therefore, yield coefficient (Y) is defined as the amount of biomass produced per amount of BOD removed.

[1] *Reference*: Presentation - "Activated Sludge Process" prepared by Michigan Department of Environmental Quality, Operator Training and Certification Unit

Example 5.2

The average concentration of BOD in the entering stream to the aeration tank is 125 mg/L and the outlet BOD of the clarified effluent is 5 mg/L. Average plant flow rate is 2 MGD, while the average returned activated sludge (RAS) concentration is 8000 mg/L. Calculate the waste activated sludge produced for yield coefficient = 0.7 and 0.5 of the process.

BOD removed = 125 – 5 = 120 mg/L

Therefore per day the mg of BOD removed is = average plant flow rate × 120 = $(2)(10^6)(3.8)(120)$ = 912 kg per day

For Y=0.7, the biomass produced = 912 × 0.7 = 638.4 kg per day

For Y=0.5, the biomass produced = 912 × 0.5 = 456 kg per day

As the excess sludge is wasted from the return sludge only, therefore, MLSS in returned sludge is also 8000 mg/L.

$$\text{Therefore, WAS} = \frac{638.4 \times 10^6}{8000} = 79800 \text{ Lpd} = 21000 \text{ for Y=0.7}$$

$$\text{WAS} = \frac{456 \times 10^6}{8000} \text{ for Y=0.5}$$

Therefore, what we may conclude. Y is less implied less amount of biomass produced per amount of BOD removed. In that the wasting is basically less in order to compensate the biomass in the system as it is already less. While the situation will be reverse once, we have the high yield coefficient. In this case, we could have an ampoule amount of luxury to meet the population of microorganisms in the reactor, even though we are wasting at higher rate. Now in case of a membrane reactor system, the yield coefficient gets increased as because of retention of solids within the system. Another aspect of the activated sludge system is the aeration rate. The oxygen required capacity (ORC) shows by the below equation[2] is the real amount of oxygen required by the system and it depends on the peal oxygen capacity required (ORCp) in kg/h.

$$\text{OCR} = \frac{\text{OCR}_p}{\alpha}\left(\frac{C_{10}}{\beta C_{20} - C_{VM}}\right) \qquad (5.4)$$

Where, OCR = oxygenation capacity required (kg/h) C_{10} = Dissolved oxygen saturation level at 10°C (≈11 mg/L) C_{20} = DO saturation level at 20°C in clean water (≈9 mg/L) CVM = Dissolved oxygen concentration reference (≈0.5 mg/L) α =ratio of efficiency in the transfer of oxygen from mixed liquor and pure water. β = DO saturation correction in mixed liquor (= 0.95). In general it is seen that there is a significant decrease in the oxygen requirement with increasing solids concentration. Now let us explain the situation. Before solving out the issue, let us raise some questions.

- Why do we need oxygen in the system?
- What is 'solid' and how they are formed?
- And how the solid generation is related to the oxygen requirement?

Before attending the first question let us go with the second one. In activated sludge process wastewater containing organic matter goes into an aeration tank, micro-organisms metabolize the suspended and soluble organic matter. Portion of this organic matter is utilized towards the formation of new cells and the rest is oxidized into CO_2 and water to derive energy. These new cells formed in the reaction are separated from the liquid stream in the form of a flocculent sludge in settling tanks. Portion of this settled biomass is called activated sludge and is returned to the aeration tank, while the remaining is removed as waste or excess sludge. This concentration of biomass is our solid and more solids will be

[2] *Reference*: "Membrane Bioreactors (MBR): Series: Secondary Treatments" in "Technology Fact Sheets for Effluent Treatment Plants on Textile Industry" (FS-BIO-010) by Universidade Da Coruña

there once more and more cells will be formed. Therefore, in order to limit the solid concentration, we must go for reduced aeration *i.e.*, oxygen supply.

Table 5.1: Value of oxygen transfer coefficients at different MLSS concentration

MLSS concentration (mg/L)	*Oxygen transfer coefficient*
3,000-6,000	0.6–0.8
10,000-12,000	0.5–0.6
15,000-20,000	0.4

5.3. CHAPTER IN BRIEF

- Conceptual undertsanding of membrane reactor.
- Understanding Activated Sludge Process.
- Ideas on different types of membrane reactor used in wastewater treatment.
- Submerged membrane reactor design.
- Concept of MLSS, oxygen required in the aeration tank and oxygen uptake.

Chapter 6

Basics of Numerical Methods and Stochastic Simulation

This chapter deals with the basics of numerical methods that are useful in the calculation of the model equations for membrane separation process. Now the foremost question that comes up in our mind is how do you define an equation? Or more specifically from an engineering point of view what is process model? Now let us say a parameter 'y' is depending on x_1, x_2, x_3,.....,x_n. Then the functional relationship can be given as

$$y = f(x_1, x_2, x_3, \ldots\ldots, x_n) \tag{6.1}$$

Where, $x_n \in R$ belongs to real numbers.

Now equation 6.1 equality can only be valid once the relationship in equation 6.2 holds good for any value of x_i, i=1,2,3,...,n

$$\psi(x_1, x_2, x_3, \ldots\ldots, x_n, y) = y - f(x_1, x_2, x_3, \ldots\ldots, x_n) = 0 \tag{6.2}$$

This is what we are calling *equation*. And to be more specific this is what the *model* also. So in brief, the process model is an equation that expresses the relationship among the process variables. Now, the model can be of two types – one is at the steady state and the other is at the unsteady state. Again, both the models be bifurcated into another two divisions – one is the linear model and the other one is the non-linear model. This chapter deals with those mathematical approaches that are pertinent to the membrane separation process. The main topics that are discussed here is the matrix manipulation, solution of linear and nonlinear equations, solution to ordinary differential equations and solution to partial differential equations. In the second part we will discuss the stochastic process and the solution to a stochastic process, which is much pertinent to the membrane reactor or to the normal membrane separation process.

6.1. SOLUTION TO SIMULTANEOUS LINEAR EQUATION

The simultaneous equations are those, where several dependent parameters individually are depending on a particular set of independent parameters and also the individual correlation holds good. When the dependency is with the order of one with respect to the independent parameter, then we are calling it simultaneous linear equations. So in general we can write the formulation as below.

$$y_i = \sum_{j=1}^{n} a_{ij} x_j \text{ for } i=1,2,3,\ldots,N \qquad (6.3)$$

This can also be written as Y=Ax, where A is coefficient matrix and X is called the solution matrix. To obtain the solution the determinant of coefficient matrix could not be zero and the matrix can be solved through Cramer's rule. Now again to solve this matrix we need to understand the degrees of freedom, which is given by the difference between the number of equations and the number of variables. For the equation 6.3 the equation can be re-arranged as

$$y_i - \sum_{j=1}^{n} a_{ij} x_j = 0 = f\left(x \in \{x_i\}, y \in \{y_i\}\right) \qquad (6.4)$$

Now the number of equations here is equal to 'N', when the number of variables comes as 'n+N'. Therefore, the degrees of freedom (dof) = (n+N) – N = n. Therefore, the unique solution of the system cannot be found. In that case we need to go for the fixation of 'n' variables to make the degrees of freedom is equal to zero and have a unique solution to the system. The system is called *overdetermined*, when N>n and the reverse is called *underdetermined.* In this chapter we will discuss on Jacobi iteration method to solve the problem, although we have several other methods like Gauss elimination, Gauss Jordon and Gauss-Siedel method. The condition for this is that the matrix must be diagonally dominant.

Jacobi is an iterative method, where the solution starts with an initial guess of the variables. Let us start with an example for a unique solution system.

$$\left.\begin{aligned} f(x_1,x_2,y_1) = y_1 - (a_{11}x_1 + a_{12}x_2) \\ f(x_1,x_2,y_2) = y_2 - (a_{21}x_1 + a_{22}x_2) \end{aligned}\right\} \tag{6.5}$$

Now, as dof (= 4 – 2) = 2, hence to get unique solution of the system we need to fix two of the variables to get the solution. Therefore, in order to reduce the dof, we will make constants as y_1 and y_2. Therefore, the relations 6.4 can be said as,

$$\left.\begin{aligned} (a_{11}x_1 + a_{12}x_2) = b_1 \\ (a_{21}x_1 + a_{22}x_2) = b_2 \end{aligned}\right\} \tag{6.6}$$

To solve for equation 6.6 one can write,

$$x_1 = \frac{b_1 - a_{12}x_2}{a_{11}} \tag{6.7a}$$

$$x_2 = \frac{b_2 - a_{21}x_1}{a_{22}} \tag{6.7b}$$

In general equation 6.7 can be written as,

$$x_i = \frac{b_i - \sum\limits_{\substack{j=1 \\ \neq i}}^{n} a_{ij}x_j}{a_{ii}} \tag{6.8}$$

Now in Jacobi method the first two conditions seen are – one the determinant of coefficient matrix is not equal to zero and the matrix is diagonally dominant. It is an iterative method and the algorithm is given by the equation 6.9.

$$x_i^{new} = \frac{b_i - \left(\sum_{j=1}^{i-1} a_{ij} x_j^{new} + \sum_{j=i+1}^{n} a_{ij} x_j^{old} \right)}{a_{ii}} \tag{6.9}$$

Initially all the values for x's will be assigned with initial guess value. Just take equations 6.7a and b. Initially with the guess value of x_2, x_1 is calculated. Once the value for x_1 is calculated, with the value x_2 is evaluated. Suppose in this case, we have four variables x_1, x_2, x_3 and x_4. To calculate the value for x_1, all the old values for x_2 to x_4 were evaluated. To calculate the value for x_2, we will take the newly evaluated value for x_1 and rest of all the values will be the previous one. For x_3, we will take the new values for x_1 and x_2. For x_4, all the values will be the new values. This algorithm is summarized in equation 6.9. To stop the iteration, we will set the convergence criterion as $\left| x_i^{new} - x_i^{old} \right| \le \varepsilon$, where 'ε' is a chosen value close to zero called *tolerance*.

6.2. SOLUTION TO SIMULTANEOUS NONLINEAR EQUATION

In the previous section we have discussed the solution to simultaneous linear equations. However, most of the equations or models are nonlinear only. In this section we will discuss Newton-Raphson method, the most common one for the solution of nonlinear equations. It is a method following the gradient approach to get the solution of a nonlinear equation. Let us say we have a nonlinear equation like $f(x) = x^n - a = 0$ and we want to find a root around a point x_1. Then according to Newton-Raphson the formulation can be said as,

$$x^{new} = x^{old} - \frac{f(x)\big|_{x=x^{old}}}{\frac{df(x)}{dx}\Big|_{x=x^{old}}} \tag{6.10}$$

Fig. 6.1 will elaborate the process and the iteration will go on until the condition $\left|x^{new} - x^{old}\right| \le \varepsilon$ is satisfied.

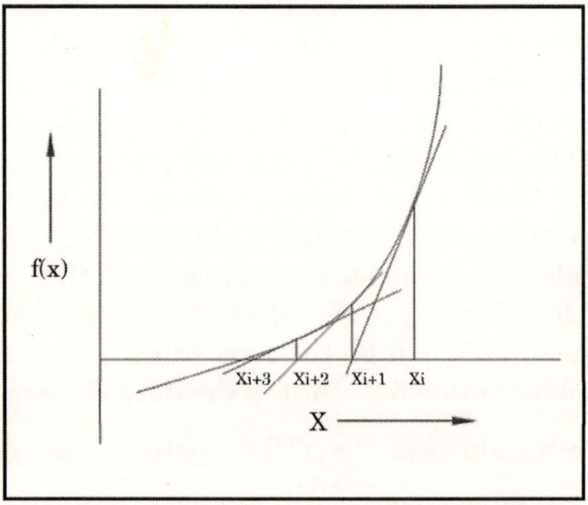

Fig. 6.1: Graphical representation of Newton-Raphson method

Now the discussion is with one variable only. If there will be more than one variable, then how the method is going to be implemented? Let us go with nonlinear functions as given below.

$$\left.\begin{aligned} &f_1(x_1, x_2, ..., x_n) = 0 \\ &f_2(x_1, x_2, ..., x_n) = 0 \\ &\vdots \\ &f_n(x_1, x_2, ..., x_n) = 0 \end{aligned}\right\} \tag{6.11}$$

The derivative of the function is called the Jacobian (J) and the new set of values will be evaluated according to the Newton's method algorithm. Equation 6.12 gives the algorithm.

$$\left\{x_i^{new}\right\} = \left\{x_i^{old}\right\} - J^{-1}\left\{f_i^{old}\right\}, \text{ where } \quad J = \begin{bmatrix} \frac{\partial f_1}{\partial x_1} & \frac{\partial f_1}{\partial x_2} & \cdots & \frac{\partial f_1}{\partial x_n} \\ \frac{\partial f_2}{\partial x_1} & \frac{\partial f_2}{\partial x_2} & \cdots & \frac{\partial f_2}{\partial x_n} \\ \vdots & \vdots & \vdots & \vdots \\ \frac{\partial f_n}{\partial x_1} & \frac{\partial f_n}{\partial x_2} & \cdots & \frac{\partial f_n}{\partial x_n} \end{bmatrix} \quad (6.12)$$

6.3. ORDINARY DIFFERENTIAL EQUATION

In general any of the processes vary with different parameters and if the process is a continuous one then the whole model will be evaluated in the form of differential equation. Now in case of ordinary differential equation (ODE) the dependency lies on only one type of the variable. Say during ultrafiltration process we have the accumulation over the membrane and the deposition increases with time. Therefore, if we want to go for mass balance over the membrane surface we have two dependencies – one, with time the concentration over the membrane surface will vary and the second one is the distance from the surface of the membrane. Under steady state condition, we have only variation of the concentration with the distance from the membrane and the differential function looks like $f(x) = \dfrac{d\psi(x)}{dx}$, where $\psi(x)$ is the concentration term. After the solution of ODE, we will have a profile of the original function (here $\psi(x)$) with 'x', the independent parameter. There are different methods by which one evaluate and thereafter understand the profile for the function. The preliminary one is called the Euler's method and the algorithm comes from the concept of Taylor's expansion of a function around any point. Now, what we understand from this Taylor's expansion? We understand the nature of functional variation with respect to the independent parameter.

From the said elaboration here if the function $\psi(x)$ will be expanded around a point x_0, then the Taylor's expansion can be written as,

$$\psi(x)=\psi(x_0)+(x-x_0)\frac{d\psi(x_0)}{dx}+\frac{(x-x_0)^2}{2!}\frac{d^2\psi(x_0)}{dx^2}+\ldots\ldots \quad (6.13)$$

Now, if we are neglecting the second order term from equation 6.13, then it becomes,

$$\psi(x)=\psi(x_0)+(x-x_0)\frac{d\psi(x_0)}{dx}+O(h^2),\ \text{where } h=x-x_0 \quad (6.14)$$

This is what the Euler's method algorithm.

$$\psi(x_{i+1})=\psi(x_i)+hf(x_i) \quad (6.15)$$

Now the profile is available once we have the value for the function $\psi(x)$ at $x=x_0$. This type of problem is called *initial value problem (IVP)*.

However, this method has a limitation in predicting the process. There is another method called Heun's method, which is a combination of both explicit and implicit method. The algorithm is given below.

$$\psi^{P}(x_{i+1})=\psi(x_i)+hf(x_i,\psi_i)\ \text{(Predictor)}$$
$$\psi(x_{i+1})=\psi(x_i)+\frac{h}{2}\left[f(x_i,\psi_i)+f(x_i,\psi^{P}_{i+1})\right]\ \text{(Corrector)} \quad (6.16a)$$

The most appropriate method for solving ODE is the Runge-Kutta method, which encompasses all the methods that are used for solving ODEs. Euler's method is the first order Runge-Kutta method. The below algorithm described the 4th order Runge-Kutta method.

$$
\begin{aligned}
k_1 &= f(\psi_i, x_i) \\
k_2 &= f\left(\psi_i + {}^{k_1}\!/_2, x_i + {}^{h}\!/_2\right) \\
k_3 &= f\left(\psi_i + {}^{k_2}\!/_2, x_i + {}^{h}\!/_2\right) \\
k_4 &= f(\psi_i + k_3, x_i + h) \\
x_{i+1} &= x_i + \frac{h}{6}(k_1 + 2k_2 + 2k_3 + k_4)
\end{aligned}
\tag{6.16b}
$$

6.4. PARTIAL DIFFERENTIAL EQUATION

When a dependent parameter is depending on several independent parameters in a continuous domain, then its tracking will be depending individually on the changes of each of the independent parameters. The changes will be expressed as the differential equation(s), which is called *partial differential equation (PDE)*. The PDE can be written as,

$$
f(x_1, x_2, x_3, \ldots, x_n) = \frac{\partial f}{\partial x_1}\Delta x_1 + \frac{\partial f}{\partial x_2}\Delta x_2 + \frac{\partial f}{\partial x_3}\Delta x_3 + \ldots + \frac{\partial f}{\partial x_n}\Delta x_n \tag{6.17}
$$

This is what a very common in mass balance equation against a membrane. The accumulation term is the variation of concentration with time, while the convective term is the spatial variation of concentration over the membrane within selected domain within the mass transfer boundary layer. The equation looks like,

$$
\frac{\partial f}{\partial t} = k\frac{\partial^2 f}{\partial x^2} \tag{6.18}
$$

This type of equation is called *parabolic equation*. When the accumulation term does not exist and the special variation of a property is multidirectional then the equation takes the form

$\frac{\partial^2 f}{\partial x^2}+\frac{\partial^2 f}{\partial y^2}+\frac{\partial^2 f}{\partial z^2}=g(x,y,z)$, which is called *elliptic equation*. In case of any oscillatory system, the equation will be $\frac{\partial^2 f}{\partial t^2}=k\frac{\partial^2 f}{\partial x^2}$, which is called *hyperbolic equation*. Now, let us go with the equation 6.18 and try to learn how to solve the form. If we draw a specified domain for the equation 6.18, it looks like shown in Fig. 6.2.

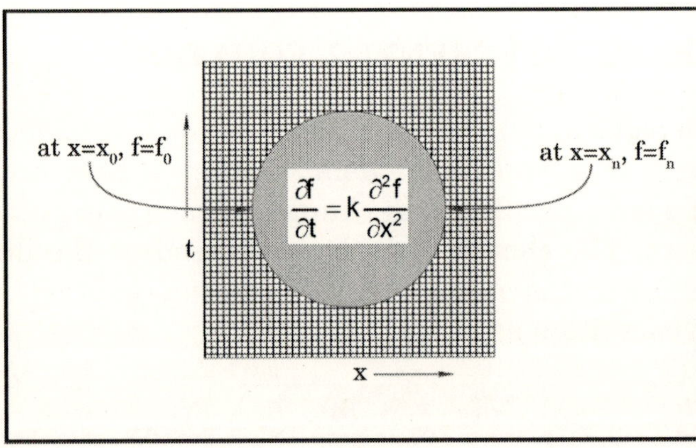

Fig. 6.2: Representation of domain for the parabolic equation

In Fig. 6.2, the x-axis represents the spatial gridding, while the y-axis represents the time steps. The boundary conditions are given as at $x=x_0$, $f=f_0$ and at the end boundary $x=x_n$, $f=f_n$ for any $t>0$. At $t=0$, $x\geq 0$ $f=f_c$.

Now there is a scheme, which is called forward difference scheme, through which we can express the differential equation to a simple algebraic equation.

$$\frac{\partial f}{\partial t}=\frac{f^{i,j+1}-f^{i,j}}{\Delta t} \tag{6.19}$$

$$\frac{\partial^2 f}{\partial x^2}=\frac{f^{i-1,j+1}-2f^{i,j+1}+f^{i+1,j+1}}{\Delta x^2} \tag{6.20a}$$

$$\frac{\partial^2 f}{\partial x^2} = \frac{f^{i-1,j} - 2f^{i,j} + f^{i+1,j}}{\Delta x^2} \tag{6.20b}$$

Equation 6.20a shows the difference in function based on space domain at the elevated time scale and 6.20b shows the difference in function at the current time scale. Now, combining equation 6.19 and 6.20a we obtain,

$$\begin{aligned} &\frac{f^{i,j+1} - f^{i,j}}{\Delta t} = k\frac{f^{i-1,j+1} - 2f^{i,j+1} + f^{i+1,j+1}}{\Delta x^2} \\ &\Rightarrow -\lambda f^{i-1,j+1} + (1+2\lambda)f^{i,j+1} - \lambda f^{i+1,j+1} = f^{i,j} \end{aligned} \tag{6.21}$$

The equation describe in 6.21 is basically the *implicit forward difference scheme*. Again combining equation 6.19 and 6.20b, we obtain,

$$\begin{aligned} &\frac{f^{i,j+1} - f^{i,j}}{\Delta t} = k\frac{f^{i-1,j} - 2f^{i,j} + f^{i+1,j}}{\Delta x^2} \\ &\Rightarrow f^{i,j+1} = \lambda f^{i-1,j} + (1-2\lambda)f^{i,j} + \lambda f^{i+1,j} \end{aligned} \tag{6.22}$$

Equation 6.22 is called the *explicit forward difference scheme*. Here, the gridding or the segmentation of the axis depends on the value of 'λ', which is given as, $\lambda = \frac{k\Delta t}{\Delta x^2}$. To maintain the stability of the solution the value of λ should be less than or equal to 0.5.

Now for the problem given in Fig. 6.2 if the origin of the ordinates will be (0,0) (node number then the 1st and last node equation will consists of the boundary values. This type of problem is called *boundary value problem (BVP)*. The equation at node 1 from equation 6.21 can be said as,

$$\begin{aligned} &-\lambda f^{0,1} + (1+2\lambda)f^{1,1} - \lambda f^{2,1} = f^{1,0} \\ &\Rightarrow (1+2\lambda)f^{1,1} - \lambda f^{2,1} = f_c + \lambda f_0 \end{aligned} \tag{6.23}$$

For the next node it will be $-\lambda f^{1,1} + (1+2\lambda) f^{2,1} - \lambda f^{3,1} = f_c$ and for the last node it must be $-\lambda f^{n-1,1} + (1+2\lambda) f^{n,1} = f_c + \lambda f_n$. Therefore, we will have a tridiagonal matrix in the format,

$$\begin{bmatrix} (1+2\lambda) & -\lambda & 0 & 0 & \dots & 0 & 0 \\ -\lambda & (1+2\lambda) & -\lambda & 0 & \dots & 0 & 0 \\ 0 & -\lambda & (1+2\lambda) & -\lambda & \dots & 0 & 0 \\ \vdots & \vdots & \vdots & \vdots & \vdots & \vdots & \vdots \\ 0 & 0 & 0 & 0 & \dots & -\lambda & (1+2\lambda) \end{bmatrix} \begin{bmatrix} f^{1,1} \\ f^{2,1} \\ f^{3,1} \\ \vdots \\ f^{n,1} \end{bmatrix} = \begin{bmatrix} f_c + \lambda f_0 \\ f_c \\ f_c \\ \vdots \\ f_c + \lambda f_n \end{bmatrix} \tag{6.24}$$

This matrix is called *tridiagonal matrix array* and is limited to the implicit scheme. However, the explicit one is straight forward and no need to form a matrix for its manipulation. Now, there is an algorithm called Thomas algorithm through which one can solve the system and the algorithm is applicable to diagonally dominant matrix. Now according to the algorithm, the matrix is converted to upper triangular matrix. Say, we have 6x6 matrix which is tridiagonal as per the following format.

$$\begin{bmatrix} b_1 & c_1 & 0 & 0 & 0 & 0 \\ a_2 & b_2 & c_2 & 0 & 0 & 0 \\ 0 & a_3 & b_3 & c_3 & 0 & 0 \\ 0 & 0 & a_4 & b_4 & c_4 & 0 \\ 0 & 0 & 0 & a_5 & b_5 & c_5 \\ 0 & 0 & 0 & 0 & a_6 & b_6 \end{bmatrix} \begin{bmatrix} x_1 \\ x_2 \\ x_3 \\ x_4 \\ x_5 \\ x_6 \end{bmatrix} = \begin{bmatrix} r_1 \\ r_2 \\ r_3 \\ r_4 \\ r_5 \\ r_6 \end{bmatrix} \tag{6.25}$$

Now if we form one equation like this,

$$1 + \frac{c_1}{b_1} x_1 = \frac{r_1}{b_1} \Rightarrow 1 + \alpha_1 x_1 = \beta_1 \tag{6.26}$$

Again after mathematical manipulation for the second row, we can write, $x_2 + \alpha_2 x_3 = \beta_2$, where $\alpha_2 = \dfrac{c_2}{b_2 - a_2\alpha_1}; \beta_2 = \dfrac{r_2 - a_2\beta_1}{b_2 - a_2\alpha_1}$.

Similarly for the third row, $x_3 + \alpha_3 x_4 = \beta_3$, where $\alpha_3 = \dfrac{c_3}{b_3 - a_3\alpha_2}; \beta_3 = \dfrac{r_3 - a_3\beta_2}{b_3 - a_3\alpha_2}$. The last row will be having $\beta_6 = \dfrac{r_6 - a_6\beta_5}{b_6 - a_6\alpha_5}$. The new matrix looks like,

$$\begin{bmatrix} 1 & \alpha_1 & 0 & 0 & 0 & 0 \\ 0 & 1 & \alpha_2 & 0 & 0 & 0 \\ 0 & 0 & 1 & \alpha_3 & 0 & 0 \\ 0 & 0 & 0 & 1 & \alpha_4 & 0 \\ 0 & 0 & 0 & 0 & 1 & \alpha_5 \\ 0 & 0 & 0 & 0 & 0 & 1 \end{bmatrix} \begin{bmatrix} x_1 \\ x_2 \\ x_3 \\ x_4 \\ x_5 \\ x_6 \end{bmatrix} = \begin{bmatrix} \beta_1 \\ \beta_2 \\ \beta_3 \\ \beta_4 \\ \beta_5 \\ \beta_6 \end{bmatrix} \tag{6.27}$$

Therefore in general the algorithm can be written as (excluding for the first and the last row),

$$\alpha_i = \frac{c_i}{b_i - a_i\alpha_{i-1}}; \beta_i = \frac{r_i - a_i\beta_{i-1}}{b_i - a_i\alpha_{i-1}}.$$

The solution can be written as, $x_n = \beta_n; x_{n-1} = \beta_{n-1} - \alpha_{n-1}x_n$. Once you will evaluate the function for $f^{i,1}$, you can go for the calculation of property value for the subsequent time steps. Now during the simulation at a particular time step all the space nodes will be calculated. Once all the space nodes are evaluated next time step will be evaluated again for the individual space node.

6.5. NUMERICAL INTEGRATION

We have studied so far ODE and PDE, which gives us the gradient of a parameter. Now if we want to get some absolute value of the parameter, we need to go for the integration of the parameter with respect to the independent one to have the cumulative value of the dependent within the entire domain of the independent one. There are different algorithms that we may apply for obtaining the value. However, here we will discuss the mostly discussed algorithm called *Simpson's rule*. The primary issue during the integration procedure is the domain segment length after dividing the entire into small segments. The length might be different for different divisions. Or in a different way depending on the variation of the function, we may say that we can change the segmentations. This is called adaptive Simpson's algorithm. The algorithm for fixed difference is given as

$$\int_a^b f(x)dx = \frac{\Delta x}{3}\left[f(x_0) + 4\{f(x_1) + f(x_3) + f(x_5) + \ldots\} + 2\{f(x_2) + f(x_4) + f(x_6) + \ldots\} + f(x_n)\right]$$

$$= \frac{\Delta x}{3}\left[f(a) + 4\{f(x_1) + f(x_3) + f(x_5) + \ldots\} + 2\{f(x_2) + f(x_4) + f(x_6) + \ldots\} + f(b)\right] \tag{6.28}$$

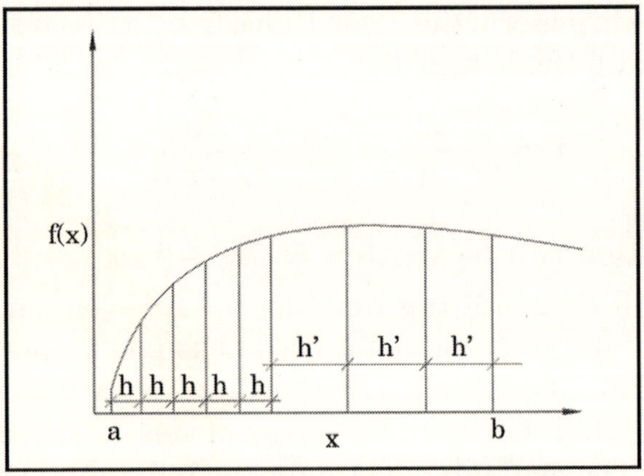

Fig. 6.3: The segregation of the area under the curve by function f(x) for the entire domain (a,b) with different step size

In the above expression we have used fixed length. But sometimes this is redundant to go for a fixed length differences for the entire domain. Let us go with the Fig. 6.3, where the first part of the function has substantial variation in the gradient, while the second part is not varying. In the adaptive process, the domain is first divided into two subdomains. Then we will be deciding whether each of these two subdomains will be further divided into more subintervals or not. This recursive process will be continued until the specified accuracy is ensured.

$$\int_a^b f(x)dx = \underbrace{I(a,b)}_{\substack{\text{Integration} \\ \left(\substack{\text{Area under} \\ \text{the curve}}\right)}} + \underbrace{E(a,b)}_{\text{Error}} \tag{6.29}$$

Now the integration part I(a,b) can be written as given in equation 6.28. Only the difference here that I would suggest is instead of diving the domain, take the whole domain with an intermediate node c, which is the arithmetic mean of a and b.

Therefore, equation 6.28 reduces to,

$$\begin{aligned} I(a,b) &= \frac{c-a}{3}\left[f(a)+4\{f(c)\}+f(b)\right] \\ &= \frac{b-c}{3}\left[f(a)+4\{f(c)\}+f(b)\right] \\ &= \frac{b-a}{6}\left[f(a)+4\left\{f\left(\frac{a+b}{2}\right)\right\}+f(b)\right] \end{aligned} \tag{6.30}$$

The error term is $E(a,b) = -\frac{1}{90}\left(\frac{b-a}{2}\right)^5 f^4$ (6.31)

Now say for the first iteration, $\int_a^b f(x)dx = I^1(a,b)+E^1(a,b)$, where the entire range h=b-a. For the second iteration, h=c-a=b-c. Therefore, $I^2=I^2(a,c)+I^2(c,b)$ and we can also write, $\int_a^b f(x)dx = I^2(a,b)+E^2(a,b)$. Now subtracting 1st iteration from the second one, one can say,

$$\begin{aligned} &I^2(a,b)+E^2(a,b)=I^1(a,b)+E^1(a,b) \\ &\Rightarrow I^2(a,b)-I^1(a,b)=E^1(a,b)-E^2(a,b) \end{aligned} \tag{6.32}$$

Again,

$$E^2(a,b)=E(a,c)+E(c,b)=E(a,b)=-\frac{1}{90}\left(\frac{h/2}{2}\right)^5 f^4-\frac{1}{90}\left(\frac{h/2}{2}\right)^5 f^4=\frac{1}{16}E^1(a,b) \tag{6.33}$$

Therefore, equation 6.32 becomes,

$$I^2(a,b)-I^1(a,b)=E^1(a,b)-E^2(a,b)=15E^2(a,b) \tag{6.34}$$

Again the integration becomes,

$$\int_a^b f(x)dx=I^2(a,b)+\frac{1}{15}\left(I^2(a,b)-I^1(a,b)\right) \text{[1]} \tag{6.35}$$

$$\frac{1}{15}\left|I^2(a,b)-I^1(a,b)\right|<\text{tolerance} \tag{6.36}$$

The iteration will stop, once $\frac{1}{15}\left|I^2(a,b)-I^1(a,b)\right|<\text{tolerance}$. If the tolerance will not be met, we will subdivide the subdomains again and iterate the process.

6.6. STOCHASTIC SIMULATION

Stochastic process is something, which depends on the probability of its happening. Therefore, once we will go with this type of process, the success of predicting the process depends on the success of predicting the events' probability. For example, say we

[1] $I^1(a,b)=\frac{h}{6}\left[f(a)+4f(c)+f(b)\right]$; $I^2(a,b)=\frac{h}{12}\left[f(a)+4f(d)+2f(c)+4f(e)+f(b)\right]$

want to understand the fouling resistance of the membrane after an hour of the separation process. Now, the membrane has some interaction with the materials need to be separated using the membrane and this depends on the retention of the materials on the surface. Therefore, the prediction on how much deposition will be there is difficult to assess. Now if one can enumerate the process with utmost accuracy after successful time based evaluation of process principles with some random variables, then such mathematical approach is called *stochastic simulation*. Fig. 6.4 gives an idea on the stochastic simulation.

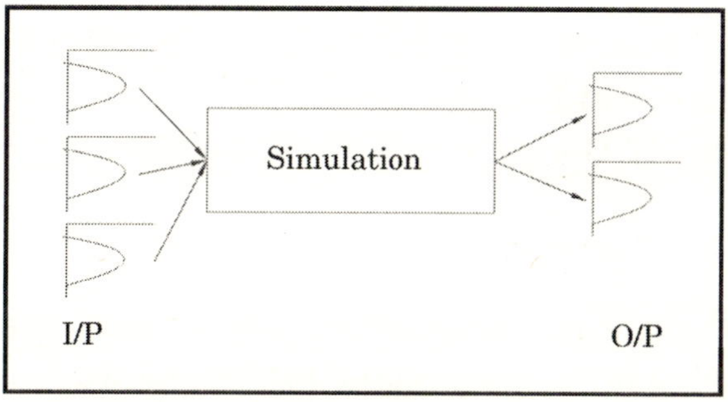

Fig. 6.4: Stochastic simulation

The basic mass transfer equation, which is called *convection-diffusion equation*, is given as,

$$\partial c_i + u.\nabla c_i = D\nabla^2 c_i \tag{6.37}$$

Now as we said that the retention of the solutes on the mass transfer is primarily depending on the possibility of staying the solutes on the surface. Therefore, the net process becomes stochastic and hence, it is quite obvious that the representation of the differential equation in the stochastic format will give us an advanced chance of accuracy in the prediction. The process can be said as the *Wiener Process*, named after the name of Norbert Wiener, a time scaled stochastic process. Often, we are calling this process is the random walk process. It is said that any process following the Wiener one, is a random variable W(t) within the interval [0,T] and depends continuously on $t\in[0,T]$, where T is the time period. At time t=0, the walk is restricted and therefore, W(0)=0 and $dW - N(0,1)\sqrt{dt}$, where N(0,1) is the Gaussian normal distribution with variance is equal to 1.

The differential equation for any process X(t) in compliance to the equation 6.37 can be written as (Szymczak and Ladd, 2004),

$$dX + v(x)dt = \sqrt{2D}dW \tag{6.39}$$

We have selected a domain called '∀' is bounded by the surface 'S', and a system of units is considered in a way like the root-mean-square displacement in unit time given by $\sqrt{2D}$ is unity. Therefore, after Δt time step the state of the process is given by,

$$X(t+\Delta t) = X(t) + v(X(t))\Delta t + \sqrt{2D}\Delta W(t) \tag{6.40}$$

This is what the point of solution, where the state of the process is approximated with infinitesimal time steps and the distribution of W is considered as the normal distribution as said earlier. If you just look at the equation 6.40, it says about the effect of both velocity and random walk on the state of the process. If Δt=0, the velocity effect will be immediately nullified, while the effect of random distribution will also be zero according to the property of *Wiener process*. Hence, the state remains same as the previous one.

6.7. CHAPTER IN BRIEF

- Different numerical techniques to model a system, solve a system and their related algorithms.
- Stochastic process and simulation.

Chapter 7

Mathematical Models for Membrane Separation Process

So far we had discussed many applications of membrane separation processes some of which a bit intricate and require rigorous analysis to optimize the process. For example, membrane reactor, where during the production of GOS so much deposition on the membrane surface limits the reaction. Also as you had seen that even during the production of COS with a dual reactor system, the reactant flow rate plays an important role to the yield of reaction. Therefore, it is important if we can develop some mathematical relationship for the membrane separation process, it will help to build a scaled up design for the unit that can be effectively implemented. In this chapter, we will discuss from a very basic model for membrane separation process to an advanced application specific model.

7.1. FILM MODEL

During the solute passage through the membrane, the solute molecules are getting deposited on the surface of the membrane and acts as a secondary barrier to the passage of the low molecular weight solutes through the membrane. This is called concentration polarization over the membrane and this is some unique feature of a filtration process that shifts the driving force from pressure difference across the membrane to mass difference. Fig. 7.1 shows a schematic representation of concentration polarization. If the solute concentration within the layer close to the membrane is given by C_m, and the solute concentration in the bulk is C_b, then they are related $C_m=cC_b$. 'c' is called the coefficient of polarization.

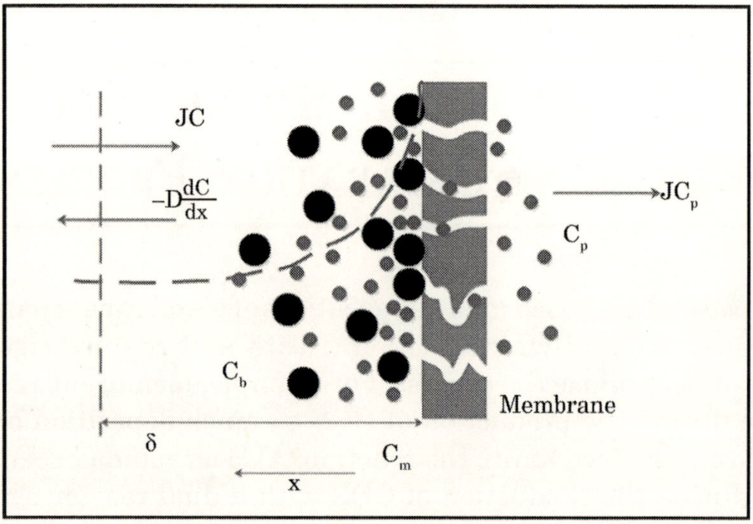

Fig. 7.1: Schematic view of permeation of solute through the membrane

At steady state, the solute balance is written:

$$J_s = JC_p = JC - D\frac{dC}{dx} \tag{7.1}$$

Where, J_s is the steady state flux, JC_p is the convective flux of low molecular weight solutes through the membrane, C_p is the concentration of the solute in the permeate and $J_d = -D\frac{dC}{dx}$ is the back-diffusional flux taking the solutes away from the membrane. D is the solute diffusivity in the solvent, which may vary with the concentration of the solute. Say, for the time if we consider that D is constant and equation 7.1 can be rearranged as,

$$\frac{J}{D}dx = \frac{dC}{C - C_p} \tag{7.2}$$

On integration of equation 7.2 with a boundary condition, at x=0, $C=C_m$ and at x=δ, $C=C_b$, it becomes,

$$J = \left(-D/_{\delta}\right)\ln\frac{C_m - C_p}{C_b - C_p} \tag{7.3}$$

If k=-D/δ is the conductance or mass transfer coefficient depending on the system hydrodynamics conditions and the physicochemical proprieties of the solution, then equation 7.3 can be rewritten as,

$$J = k\ln\frac{C_m - C_p}{C_b - C_p} \tag{7.4}$$

The negative sign signifies that the mass transfer coefficient, which is primarily representing the convective mass transfer across the membrane and the back diffusivity are of opposite nature. The model described is called film model, which is valid after the steady state attainment of the membrane filtration process *i.e.* the growth of the mass transfer boundary layer is fully developed. The mass transfer coefficient can be calculated from a Sieder-Tate type correlation.

$$\frac{kL}{D} = Sh = Z\,Re^a\,Sc^b\left(\frac{d_h}{L}\right)^c \tag{7.5}$$

The value of Z, a, b and c depend on the boundary condition of the flow regime and the related hydrodynamics.

7.2. GEL LAYER MODEL

Now as the concentration of the solute at the membrane surface increases because of polarization, the macrosolute reaches at its solubility limit and precipitates on the membrane surface forming a solid or thixotropic gels. This will lead to a constant gel layer concentration, which is practically independent of bulk solution concentration, applied pressure, fluid flow conditions or membrane characteristics. According to gel layer model, an increase in the applied pressure across the membrane produces a temporary increase in flux, which eventually brings more solute to the gel layer and increases its thickness. This increases the hydraulic resistance to solvent flow, thereafter an ultimate flux will be reached, when 100% solute rejection could be seen. At this condition, C_p becomes zero and the equation 7.4 can be re-written as,

$$J_{lim} = k \ln \frac{C_m}{C_b} \tag{7.6}$$

J_{lim} is called the limiting flux. Although gel layer model is the most simplistic model validating most of the processes, the model also has some limitations because of the variations in the assumptions made for gel layer. For example, in case of a gel layer model, we had assumed that the solutes are being solidified at the surface of the membrane and invariant to either of the bulk concentration or the cross-flow velocity of the feed. While, in true sense this is not true. Say, we are increasing the concentration of the feed in a batch membrane separation system to a high volume concentration factor (VCF). In this case the gel layer model will be validated with probably high accuracy. On the contrary, if the bulk concentration is too low (as in the case of continuous diafiltration) the gel layer formation possibility will be minimized. Similarly if the cross-flow velocity substantiates, the deposition will be reduced and henceforth the possibility of gel layer also. Therefore, the gel layer model shows high precision, when we are doing the batch separation.

7.3. RESISTANCE IN SERIES MODEL

The basic driving force for the membrane as said in chapter 1, is the trans-membrane pressure (TMP) across the membrane surface.

This forces the low molecular weight solutes to pass through the membrane pores, if the membrane is porous. On the contrary, with the nonporous membrane (like RO), the diffusion takes place according to Fick's law. The driving force here is also TMP that induces the movement of the solutes to the skin of the membrane. Therefore, in analogous to Ohm's law the flux through the membrane can be written as a function of both driving force and the resistance as shown in equation 7.7. The diagrammatic representation of the analogy is shown in Fig. 7.2.

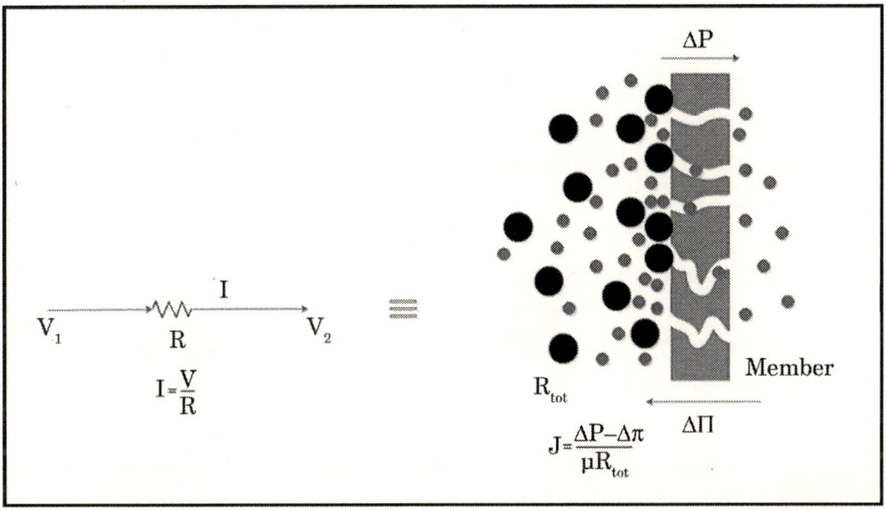

Fig. 7.2: Similarity between electrical and membrane separation process governing equation

The equation for membrane separation process is given by,

$$J = \frac{\Delta P - \Delta \pi}{\mu R_{tot}} \tag{7.7}$$

This equation is called Darcy's law, where ΔP is TMP, Δp is called the osmotic pressure acting against TMP, μ is the viscosity of the solution and R_{tot} is the total resistance acting against the flow through membrane. Now, this R_{tot} comprises of the resistance because of the membrane material (R_m), resistance because of the fouling of the membrane (irreversible) (R_f), resistance due to concentration layer (reversible) (R_{cp}) and resistance due to gel layer or cake (reversible) (R_g). Therefore, the equation 7.7 can be rewritten as,

$$J = \frac{\Delta P - \Delta\pi}{\mu\left(R_m + R_f + R_{cp} + R_g\right)} \tag{7.8}$$

However, with the application the resistance can also be modified and some extra resistance terms can also be added. Like, if we are using ultrafiltration to separate proteins from whey, the proteins get adsorbed on the surface of the membrane. So another resistance because of the adsorption of protein will also be appended with the present series resistance. However, the model shows a variation once the model will be applied to some bioreactor, where living organisms needs to be separated (Chang *et al.*, 2009). Here the after formation of the cakes, the interstices within the cake might be blocked by the presence of the low molecular weight solutes that are supposed to be moved through the membrane. But it might happen that the potency for adhering of these solutes is much higher within the cake compared to the membrane. Therefore, the cake becomes over-clogged. Like in case of activated sludge process (described in chapter 5), the membrane becomes fouled with the biomass, flocs, colloids etc. The performance of the sludge process is completely depending upon the settlement of the solids, which again depend on the physiology of the biomass, flocculants' nature etc. Therefore, the adherence of these within the cake or the membrane is also depending on the biomass nature, which might not be represented with the so called series resistance model described. It might needs some other differentiation of the resistance terms in order to get more accuracy.

7.4. SOLUTION DIFFUSION MODEL

One of the simplest ways in understanding the dynamics of

membrane separation process is to evaluate the system we are introducing to the membrane. For example, in case of osmosis the ions are rejected by the membrane that says that the membrane is almost impervious dense membrane. Or when we are dehydrating alcohol using pervaporation, where the difference in vapour pressure of a component in the feed needs to be passed through membrane and the partial pressure of the same component on the permeate side helps in the separation. In all these scenarios because of the non-porosity of the membrane, the molecules are moving across the membrane according to the Fick's law of diffusion (equation 7.9).

$$\text{Molar flux} = (\text{Diffusivity}) \times (\text{Concentration gradient}) \tag{7.9}$$

However, any mass transfer process can be expressed as a result of the driving force because of chemical potential[1]. In view of that, the component's flux across the membrane can be written as,

$$J_c = c_c V_c = c_c L_c \frac{d\mu_c}{dx} \tag{7.10}$$

Where, μ_c is the free energy of the component actively transports through the medium, L_c is called the proportionality constant and V_c is the convective velocity. Now, the change in potential is primarily the driving force because of the pressure difference across the membrane and at the same time the concentration difference. Hence, the change in chemical potential can be related as,

$$d\mu_c = RTd\ln(\gamma_c x_c) + v_c dP \tag{7.11}$$

Where, the activity coefficient γ_c comes because of the consideration of the nonideality of the solution. In case ideal solution one, γ_c=1.

Integrating the process for incompressible one equation 7.11 can be re-written as,

[1] Chemical potential is defined as the molar Gibb's free energy at constant pressure and temperature

$$\mu_c = \mu_c^0 + RT\ln(\gamma_c x_c) + v_c(P - P^0) \tag{7.12}$$

While, for the compressible gas and considering the ideal gas law,

$$d\mu_c = RTd\ln(\gamma_c x_c) + \frac{RT}{P}dP \tag{7.13}$$

Integrating equation 7.13, one can write,

$$\mu_c = \mu_c^0 + RT\ln(\gamma_c x_c) + RT\ln\left(\frac{P}{P^0}\right) \tag{7.14}$$

Now, the reference pressure P^0 can be replaced by the saturated pressure of the pure component and equation 7.12 and 7.14 can be re-written as,

$$\mu_c = \mu_c^0 + RT\ln(\gamma_c x_c) + v_c(P - P_c^s) \tag{7.15}$$

$$\mu_c = \mu_c^0 + RT\ln(\gamma_c x_c) + RT\ln\left(\frac{P}{P_c^s}\right) \tag{7.16}$$

Now, once we are going with the solution diffusion model the first assumption for the model is that the fluid on either side of the membrane is at equilibrium at the membrane interface. This assumption says implicitly that there is a continuous gradient of chemical potential across the membrane and saying that the rates of absorption and desorption at the membrane interface are much higher compared to the rate of diffusion through the membrane (Wijmans and Baker, 2006). In the second assumption is that the chemical potential expressed across the membrane is basically creating a gradient within the membrane. As the membrane is impervious, the pressure acts on the membrane is at maximum and it is uniformly distributed across the membrane. At the steady state, the concentration maintained at either side of the membrane is in equilibrium with the externals according to the first assumption. This rightly says that the solutes are continuously coming on to the membrane surface and similarly, the solutes are moving away from the surface. Therefore, with the pores' model there is a basic difference. In the solution–diffusion model it is assumed that the pressure within a membrane is uniform according to the first assumption and that there is a chemical potential gradient across the membrane according to the second assumption. On the contrary, with the pores' model, it is assumed that the permeate concentration within a membrane is uniform as within the pores the molecules are moving with the Knudsen diffusion, not following the Fick's law. Therefore, the chemical potential gradient across the membrane is uniform, while there is a gradient for pressure. Now, applying the aforesaid assumptions for the porous structured membrane, one can replace the equation 7.10 with the pressure gradient only.

$$J_c = c_c L_c v_c \frac{dP}{dx} \tag{7.17}$$

Integrating we can say,

$$J_c = k\frac{\left(P_f - P_p\right)}{l} \tag{7.18}$$

Where, P_f is the applied pressure on the feed side, P_p is the permeate side pressure, l is the thickness of the membrane and k is given as $c_c L_c v_c$. 'k' is called the Darcy's law coefficient.

On the contrary, with the solution-diffusion model equation 7.10 can be re-written as,

$$J_c = RTc_c L_c \frac{d\ln(\gamma_c x_c)}{dx} = -\frac{RTc_c L_c}{\gamma_c x_c}\frac{d(\gamma_c x_c)}{dx} \tag{7.19}$$

Now, the concentration terms can be written as,

$$c_c = M_c \rho x_c \tag{7.20}$$

Where, M_c is the molecular weight of the solute, ñ is the molar density and x_c is the mole fraction of the solute. Applying equation 7.20 in equation 7.19, we will get,

$$J_c = -\frac{RTL_c}{\gamma_c}\frac{d(\gamma_c c_c)}{dx} \tag{7.21}$$

Now saying the activity coefficient constant for the solute, we will have,

$$J_c = -RTL_c\frac{d(\gamma_c c_c)}{dx} = -D_c\frac{dc_c}{dx} \tag{7.22}$$

Where, D_c, Fick's diffusivity is given by RTL_c-.

7.5. SEMI-EMPIRICAL MEMBRANE FOULING MODEL FOR THE MEMBRANE FILTRATION PROCESS

There are different models based on the fouling mechanism have been modelled. The basis is the way of deposition on the membrane. These are like complete blocking model, intermediate blocking model, cake filtration model and standard blocking model. The first three models are primarily saying about the deposition on the membrane surface, while, the standard model says about the deposition within the pores' surface. In this section we will discuss about the first two models *viz.*, complete blocking model and intermediate blocking model. Hermia's model for cross flow filtration is the most useful and applicable models for microfiltration flux decline prediction. The model was based on the constant pressure filtration operation and based on the "intermediate blocking law". According to them, the constant pressure filtration process can be written as,

$$\frac{d^2t}{dV^2} = k\left(\frac{dt}{dV}\right)^n \tag{7.23}$$

Where, V is the volume of the permeate collected and k and n are the constants depending on the mechanism of the filtration procedure.

Again, the permeate flux collected can be written as,

$$J = \frac{1}{A}\frac{dV}{dt} \tag{7.24}$$

Where, A is the effective area of the membrane. Now combining equation 7.23 and 7.24 one can say,

$$\frac{d^2t}{dV^2} = \frac{d}{dV}\left(\frac{dt}{dV}\right) = \frac{d\left(\frac{1}{AJ}\right)}{dV} = -\frac{1}{A^2J^3}\frac{dJ}{dt} = k\left(\frac{dt}{dV}\right)^n \tag{7.25}$$

Therefore equation 7.25 can be re-written as,

$$-\frac{1}{A^2J^3}\frac{dJ}{dt} = k\left(\frac{1}{AJ}\right)^n \tag{7.26}$$

Now with the dead-end filtration and a complete blocking mechanism n=2. Therefore, equation 7.26 becomes,

$$\frac{dJ}{dt} = -kJ \tag{7.27}$$

Now, integrating the process from initial permeate flux to a permeate flux achieved after time‘t’ (when the process is mass transfer driven instead of pressure driven),

$$J = J_0 \exp(-kt) \tag{7.28}$$

Where, J_0 is the initial permeate flux. Now, according to the model, it is understood that the pore blocking is basically because of the sealing of the pores with the macrosolutes (as shown in Fig. 7.3). As the fraction of the pores is accessible to the convective flow of the microsolutes, therefore the permeate flux is equal to the fraction of the pore area opened. Hence, one can say the following relationship (Field *et al.*, 1995).

$$\frac{d\left(J/J_0\right)}{dt}=\frac{d\left(A_e/A_{e0}\right)}{dt} \tag{7.29}$$

Where, A_{e0} is the initial effective membrane area and J_0 is the initial permeate flux through the membrane.

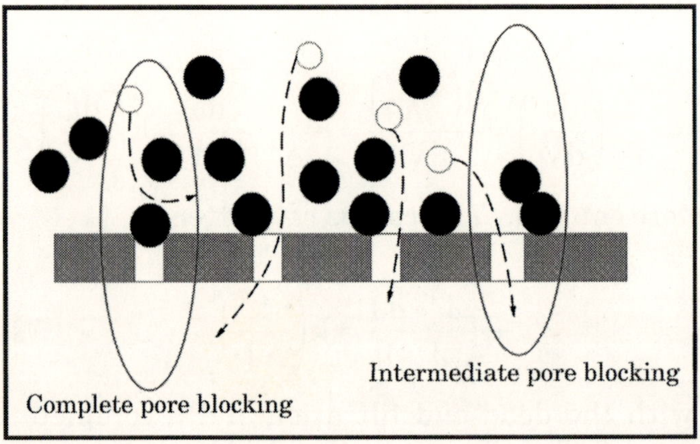

Fig. 7.3: Passage of the microsolutes through the membrane pores (dashed line show the path of the microsolutes) and sealing of the pores with the macrosolutes (black spheres are the macrosolutes)

Now, we are introducing a parameter 'a', which is the area cannot be accessed per unit volume of the filtrate. Therefore, the net effective area A_e can be written as $A_e = A_{e0} - aV$. Hence, equation 7.29 can be written as,

$$\frac{dJ}{dt}=-\left(\frac{J_0}{A_{e0}}\right)\frac{d(aV)}{dt}=-\left(\frac{aJ_0}{A_{e0}}\right)AJ \tag{7.30}$$

Now the effective area of the membrane can be written as $A_{e0} = \varepsilon A$, where 'å' is the fraction of the membrane area. Therefore, equation 7.30 can be said as,

$$\frac{dJ}{dt} = -\left(\frac{J_0}{A_{e0}}\right)\frac{d(aV)}{dt} = -\left(\frac{aJ_0}{\varepsilon}\right)J \tag{7.31}$$

Now, what the equation 7.31 is saying. It says that with the separation process moves on, the value for 'aV' is constantly increasing because of the fouling of the membrane and therefore, the pore blocking. However, after a substantial period of operation, the membrane surface gets choked and no more area to be left for further deposition. Therefore, the flux becomes asymptotically decreasing function and while, the term $\frac{d(aV)}{dt}$ is asymptotically increasing. Now, the fact is very much true, once we are going with a membrane separation process, where the fouled portion cannot be removed. What happens, when a cross-flow system is used?

In cross-flow system, the deposition is always being moved away from the surface depending on the cross-flow velocity. Therefore, the effect of pore blocking is constantly diminishing and the expected drop in the flux will be somehow being reduced. Therefore, the equation 7.29 can be re-written as,

$$\frac{d\left(J/J_0\right)}{dt} = \frac{d\left(A_e/A_{e0}\right)}{dt} + BJ_0 \tag{7.32}$$

Where, B is a constant and the term 'BJ_0' is basically giving us an idea stating that after wash the membrane has regained its flux.

Equation 7.32 will be again written as,

$$\frac{dJ}{dt} = -\left(\frac{aJ_0}{\varepsilon}\right)J + BJ_0 \tag{7.33}$$

When $J \leq \frac{B\varepsilon}{a}$, $\frac{dJ}{dt} \leq 0$ implies that there will not be any flux decline. This value of flux below or at which there will not be any further flux decline is called critical flux and given as 'J_c'. Therefore equation 7.33 becomes,

$$-\frac{dJ}{dt} = \left(\frac{aJ_0}{\varepsilon}\right)(J - J_c) \tag{7.34}$$

Hence, on integration of equation 7.34, we will get,

$$\frac{J - J_c}{J_0 - J_c} = \exp\left(-\frac{aJ_0}{\varepsilon}t\right) \tag{7.35}$$

Now in the *complete blocking* model it is assumed that each particle is blocking the open pores. On the contrary, with the *intermediate blocking* mechanism the pore blocking depends on the probability of the landing of particles on the surface and the number of open pores still there. Based on this assumption it may be said that the change in the number of pores per unit of the permeate volume collected over the unit area of the membrane in unit time is inversely proportional to the flux through the membrane. Now this what the assumption tells about. If the pores are already blocked then there should not be any new blocking exists. In this case the flux value is very low. At this point if we still continues the process, might be we have a minute change in the flux value, but the pore blocking will not be happened. Therefore, the gradient change is in the reverse direction. On the contrary if maximum pores are opened, maximum number of particles is having a tendency to clog the pores. And at the same the flux change is prompt enough along with a high value of the flux. Now as the process moves *i.e.*, with the changes in the flux, the change in effective area is getting reduced as number pore opening is getting reduced. Fig. 7.4 is providing a picture on that. Therefore, we can say,

$$\frac{dA_e}{dJ} = \frac{K}{J} \tag{7.36}$$

Where, K is a constant. On rearrangement of equation 7.36, we get,

$$\frac{1}{J}\frac{dJ}{dt} = -\frac{aA}{K}J \tag{7.37}$$

Now similar to the conception expressed for equation 7.33, we will also incorporate a constant say 'B_1' to compensate the washing of the deposition. Therefore, equation 7.37 becomes,

$$\frac{1}{J}\frac{dJ}{dt} = -\frac{aA}{K}J + B_1 \tag{7.38}$$

Saying $J_c = \dfrac{KB_1}{\sigma A}$, we can re-write the equation 7.38 as,

$$-\frac{1}{J}\frac{dJ}{dt} = \frac{aA}{K}(J - J_c) \tag{7.39}$$

Now the aforesaid derived equations 7.34 and 7.39 can be written in a generalised form like,

$$-\frac{dJ}{dt}J^{n-2} = \Psi(J - J_c) \tag{7.40}$$

Where n=2 for complete blocking and n=1 for intermediate blocking.

7.6. CHAPTER IN BRIEF

- Basic models manifest membrane separation process.
- Overview of different flux model and fouling model for the membrane separation process.
- Different types of irreversible pore blocking model.

Chapter 8

Stochastic Simulation of Membrane Separation Process

One of the difficult issues with the membrane separation process is to predict the performance of the membrane over time. Especially, difficulties shoot up, when one uses the polymeric membrane for biomolecule separation. As in case of a membrane reactor discussed in chapter 5, where the membrane behaves as a reactor along with a separation unit behaves as a medium with different interacting parameters. Such a system will be chaotic and can't be illustrated with so called mathematical idealisation. Hence, approach with stochastic modelling provides a good texture to formulate a predictive model, which can be incorporated to design a control process for membrane separation technique. In this chapter, we will discuss the applicability of stochastic simulation in membrane separation and how the technique can be deployed to develop a predictive model for membrane separation process. At the end of this chapter, an attempt will be made to bridge between the simulation model and the design of a control process for the membrane separation unit.

8.1. PENETRATION AND SURFACE RENEWAL THEORY

As said in chapter 7 the simplest models for membrane separation process proposed after consideration of the control volume within the boundary layer, where the transport is in steady state. However, in reality for a substantial period of time the permeate flux is continuously decreasing with time and hence difficult to characterize the system with the proposed steady state transport. In view of that, the separation process can be visualised as the unsteady state system following Fick's second law of mass transfer (Eq. 8.1).

$$\frac{\partial C}{\partial t} = D\frac{\partial^2 C}{\partial z^2} \tag{8.1}$$

According to the penetration theory (Higbie, 1935) a cluster of gas molecules reach at the interface of gas and liquid (say for gas absorption), when some molecules are penetration through the interface and rest of the molecules are leaving the surface. The retention time of the cluster over the interface guides the performance of the mass transfer from gas phase to liquid phase. However, as defined stochastic process arises because of randomly changed variable because of probabilities. Hence, even with the penetration theory the retention of the cluster depends chaotically on the interactivity of the gases and the solvents used during the membrane separation process. As an example in case of CO_2 absorption using membrane gas absorption process, the CO_2 rich feed gas once comes in contact with the solvent fed through membrane. The retention time of the cluster can be estimated through a distribution function by which we can introduce some adjustable parameters that tells us about the interaction between the membrane and the chemicals involved in the separation process. As said, one of the typical questions that comes up first with the concept of location of the cluster at the interface and thereafter, its dispersion in the bulk. Before, going into the deeper portion of statistical solution of the penetration theory, first discuss some of the statistical concepts in the light of penetration theory. Below are some definitions on the statistical definition with some examples.

8.1.1. Statistical Parameters

The primary keywords that have to be conceptualized first are population and sample, along with a proper identification their interrelation. Consider in a country like India, we have some around 100 crores of people, which is called the *population* of India. Now, if someone will ask, how many persons in India are suffering from type II diabetes? The answer to this question might be say around 10 lakhs people are type II diabetic. Now, can we call this 10 lakh is the *sample* as we are picking this number from the *population*? The answer should be “NO”! A *sample* can be defined as a part of the *population* that is having the same qualitative property as the *population*. For the above example, if we will pick up 100 people from different parts of Delhi (say) those who are type II diabetic, then we can say that this 100 people are the sample among the 10 lakh type II diabetic populations in India. Now, whether this sample is truly depicting the true picture of diabetic community in India or not that depends on the values estimated for different statistical parameters attributing to the interrelation between sample and population. *Mean*, *standard deviation*, *variance* and *degrees of freedom* are some of the basic parameters that are used to have the interrelation between the sample and population.

Mean is the parameter that gives the location of a population or the sample on the number line, if the sample and population, both the instances are finite and small enough. Now, the question you must ask, why the instances will be finite and small as well to find the *mean*? In statistics the number of occurrences of instances is called *frequency* and if we want to have some gross idea on the average number of occurrences of the instances we will calculate *mean*. Therefore, if the occurrences of the number of instances will be infinite, we will not have the boundaries to calculate mean. Secondly, if the occurrences of the instances will not be small enough, the occurrences will behave as a continuous one and we will calculate the *expectation* rather than *mean*.

Now the question comes – difference between the *expected value* and the *mean value*. The *expected value* is the weighted average of the occurrences and the *mean value* is the simple average. This illustrates that for sample we could have the expectation, while for population we have the actual mean. Let us illustrate one example for each to have our concept clear.

Example 8.1

A set 'S' is defined as S = (Sen *et al.* 2011,3,3,3,4,4,5). Now let us put this statement with a case study. A school called 'XY' is continuously increasing their seat capacity since 2000. In 2000, they had increased capacity by 1 no. of seat. In 2001 they had increased 2 no. of seats. In 2003 they had increased again 2 no. of seats. Thus in 2009, they had increased 5 no. of seats. Now, what is the mean value of the increase in the seats' number?

$$\text{Mean Value} = \frac{1+2+2+3+3+3+3+4+4+5}{10} = 3 \qquad \text{(E.8.1)}$$

Therefore, we can say so far on an average number of seats increased by 3. Now, the expected value of seat increase will be given as below,

Expected Value =

$$\left(1/10\right)\times 1+\left(2/10\right)\times 2+\left(4/10\right)\times 3+\left(2/10\right)\times 4+\left(1/10\right)\times 5 = 3 \quad \text{(E.8.2)}$$

This means the occurrences of '1' is of probability $1/10$, '2' is of probability $1/5$, '3' is of probability $2/5$, '4' is of probability $1/5$ and '5' is of probability $1/10$. Now, what we can understand from these two approaches. The expected value tells us that the expected yearly capacity increase will be 3 nos. with maximum probability.

Note

$$\mu\,(\text{Mean Value}) = \frac{\sum_{i=1}^{N} x_i}{N} \tag{8.2}$$

$$E(x)\,(\text{Expected Value}) = \sum xP(x) \tag{8.3}$$

Where 'x' is the occurrences and 'P(x)' is the probability of occurrences.

Standard deviation is the deviation from the mean value of the population and is written as,

$$\sigma(\text{Standard Deviation}) = \sqrt{\frac{\sum(x-\mu)^2}{\text{DOF}}} \tag{8.4}$$

Now, as said before, we can have the statistical concept with two different approaches. One, when we consider "*sample*" and the other we consider called "*population*". Once, we want to predict the outcome of a process, we are choosing samples from the population of data. These, sampling procedures are of two types – one is called "*Biased Sampling*" and the second one is called "*Unbiased Sampling*". Based on that, the standard deviation or the variance will be calculated. Moreover, the accuracy of the sample will be understood correctly, by the accuracy of the sampling, which again depends on the method of sampling.

Now before going into the details of the calculation, let us take one common life example of sampling and let us try to understand the importance of two sampling methods. Say a state is having an election in current year. The political parties, who are standing for empowerment, are called 'A', 'B' and 'C'. The basic features of party 'A' is that these people believe in equalization of the society, while 'B' and 'C' are liberal. However, 'C' is very much fascinating with the gross development of the state through industrialisation. Based on the pre-polling, it was seen that 'C' has won the majority with an almost 70% ahead of the other teams. But in actual, 'A' has won the election, with a big majority. Now the question is what was wrong with the pre-polling? Before moving into the analysis on the error made by pre-polling agency, let us define some features of the state. The state has more than 60% of the rural areas, where the average monthly income of the salary is rupees 10,000/-. While, in urban areas the average monthly income is rupees 40,000/- 50,000/-. Rural people are also reluctant to lead urban life. The pre-polling agency had mainly taken the feedback from these urban people. This means they had to consider the rural people also as the state has more than 60% rural area, because in general the less income group are fascinating the equalisation of the wealth. Therefore, the sample has to be taken more from the rural and this sampling is called the *biased sampling*, where we are compelled to choose the sampling scheme cautiously. When, we don't have such obligations during sampling, it is called *unbiased sampling*.

Now the question is what we can conclude?

Our estimates will be such that we should eliminate the possibility of biasing. In other words, we may say that our estimates must be such that it will project a deviation so that all the possibilities for deviation could be attributed. This is what has to be reflected in standard deviation calculation through DOF as shown below.

$$\sigma(\text{Standard Deviation}) = \sqrt{\frac{\sum(x-\mu)^2}{N}} \quad \text{(For Biased population)} \tag{8.5}$$

$$\sigma(\text{Standard Deviation}) = \sqrt{\frac{\sum(x-\mu)^2}{N-1}} \quad \text{(For Unbiased population)} \tag{8.6}$$

'N' is the population size

It is obvious that for biased sample we have to go with every possibility as individual incidents are parallel to each other. As in our previous example, the qualitative parameters like fascination with the development etc. are parallel. On the contrary, with the unbiased system we have dependency and if we are going to fix up some parameter that may also lead to change in another parameter. Hence, for unbiased system, we have DOF is equal to N-1.

Example 8.2:

The city council is going to vote on whether to keep its leash laws.

Sample 1: Survey dog owners

Sample 2: Survey people who don't have pets

Sample 3: Surveying some people who own dogs and some who don't.

Which sample shows the least bias?

Answer: Sample 3: Equal chances of yes or no answers

(Refer: http://www.rsm.rcs.k12.tn.us/teachers/Robertsont/6.5.3%20 Biased%20Samples%20Lesson,%20Practice,%20and%20Quiz.pdf)

8.1.2. Stochastic Simulation?

Now, one of the mostly concerned issues here is how we can relate statics and stochastic simulation. Rather, before founding any correlation or relation between these two let us discuss the basic approaches by which we can design a stochastic simulation. The word '*Simulation*' tells us about the real world scenarios with time after creating a virtual world. Like all our gaming programs. Fig. 8.1 shows a diagram, where the realistic motion of the car is tracked by a data logger and being converted to a virtual world.

The stochastic makes it more generalised as it describes the probability of occurrences with time. The mass transfer process, especially the renewal process can be approximated by a distribution called asymptotic distribution. The distribution says about the continuous occurrences of nonnegative independent events with a common distribution function as below,

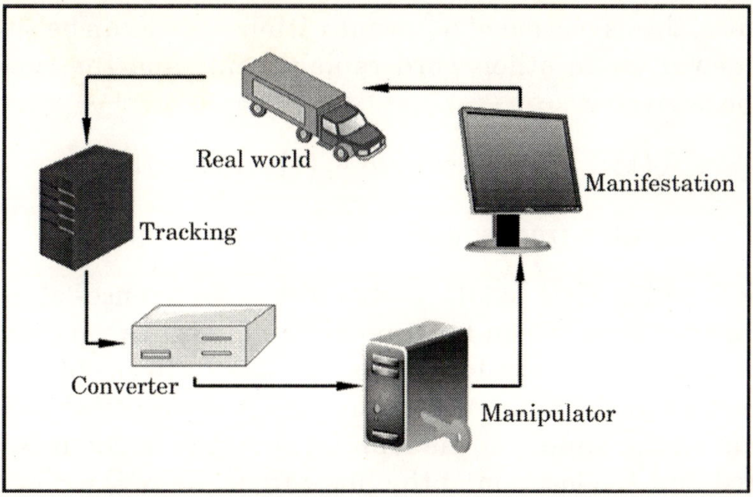

Fig. 8.1: Simulation of a car movement

$$F(x) = P\{x \geq X_k\} \;\; x \geq 0 \tag{8.7}$$

Where x is a nonnegative event and k = 1,2,3,.....Now for the distribution it can be said that $\mu = E(X_k)$ with an assumption that $0 < \mu < \infty$.

Now let us consider one example which throws some idea on the renewal process exaggerated using 'Poisson' distribution or density function. For example (Tijms, 2003), we have infinite number of bulbs' supply, where X_k burning times of bulbs are independent and identically distributed variables. If one bulb fails, it is replaced immediately. If X_i is the burning time for i^{th} bulb, then the epoch for the nth event to be executed can be given by,

$$S_n = \sum_{i=1}^{n} X_i \text{ for n=1,2,3,...} \tag{8.8}$$

Now, the total number of bulbs to be replaced upto time 't' is given by (See Fig. 8.2.)

$$N(t) = \text{The largest integer for } n \geq 0,\ t \geq 0,\ S_n \leq t \tag{8.9}$$

Hence, the execution of nth event within time 't' can be said as the renewal (or in other word replacement) and the renewal function is given as,

$$N(t) = \sum_{n=1}^{\infty} F_n(t), \text{ where } F_n(t) = P\{S_n \le t\}$$

(rewritten from equation 8.7 in terms of epoch) (8.10)

The function $F_n(t)$ *i.e.*, the probability of occurrences of the nth event within time 't' can be approximated with two parameter function. As an example, in case of the bulb problem, the replacement of the bulb (or you may call it the renewal process) depends on the supply of the bulb *i.e.*, the rate of the new bulb acquired and the location of the place from where the new bulb has been purchased and brought (Fig. 8.2).

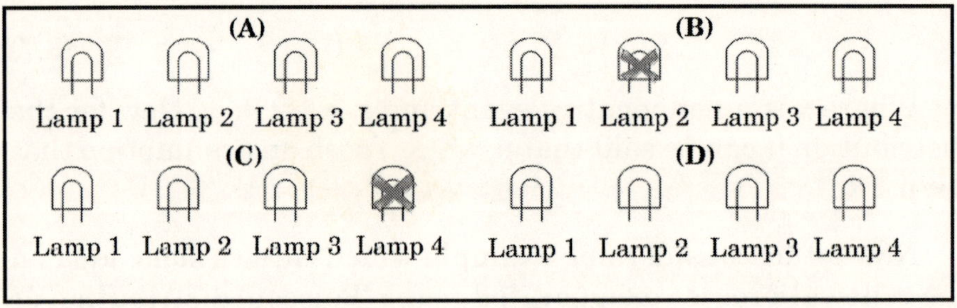

Fig. 8.2: Bulb problem

Hence, the renewal distribution can be approximated with two parameters model like Weibull distribution (called non-homogeneous Poisson process, where rate is not constant) as shown below.

$$N(t) = \sum_{n=0}^{t} F_n(t) N \int_0^t f(t)dt = \int_0^t \lambda(t) \exp\left[-\left(\frac{t}{\beta}\right)^{\alpha}\right] dt$$

$$\Rightarrow N(t) = 1 - \exp\left[-\left(\frac{t}{\beta}\right)^{\alpha}\right] \quad (8.11a)$$

Where, $$\lambda(t)=\left(\frac{\alpha}{\beta}\right)\left(\frac{t}{\beta}\right)^{\alpha-1} \tag{8.11b}$$

$\lambda(t)$ is the rate parameter, α is the shape parameter and β is the scale parameter. Now if $\alpha < 1$, then the rate parameter decreases with time 't' and *vice-versa* with $\alpha > 1$. For $\alpha = 1$, the renewal density function takes the exponential density function[1].

Example 8.3 (Abulencia and Theodore, 2015)

The life (time to failure) of a machine component has a Weibull distribution. Determine the probability that the component lasts at least 25,000 hours if

(a) The failure rate is constant and equal to 0.01 per thousand hours where t is measured in thousands of hours, and

(b) The failure rate is $t^{-1/2}$ where t is measured in thousands of hours.

Answer: $\lambda = 0.01, \alpha = -0.5, \beta = 100$

Hence, the probability of failure is given as

$$f(t)=\exp\left[-\left(\frac{25}{100}\right)^{-0.5}\right]=0.13$$

The probability of lasting, $1-0.13=0.87$

8.2. SOLUTION TO PENETRATION THEORY AND SURFACE RENEWAL THEORY USING STOCHASTIC APPROACH

As said before, penetration theory describes the retention for the packet of mass at the interface is for equal time and hence, can be

[1] The density function represents the distribution of continuous random variable, while the distribution function is the probabilistic view of the discrete variable.

approximated with the Poisson's distribution. Taking the note of the distribution function, we can say that the distribution for the packets to be created on the surface must be a variant of both time and space. Therefore, the flux near to the membrane surface can be approximated as,

Net mass flux of the micromolecules transported through the the membrane	=	Diffusion of the micromolecules from bulk towards the surface of the membrane	–	Back diffusion of the micromolecules from the membrane surface towards the bulk

(8.12)

Now the fraction of the packets or we may say the concentration has a distribution, which is exponentially varying with the distance from the membrane surface (Fig. 8.3). Now, the distribution function for concentration variation with the surface from the membrane can be written as,

$$\psi(z) = \eta \exp(-\eta z) \text{ for } z \leq \ell \tag{8.13}$$

Where, η is the distribution parameters. However, the rate of retention of the packet attributing to the concentration over the membrane can be approximated with Poisson's distribution.

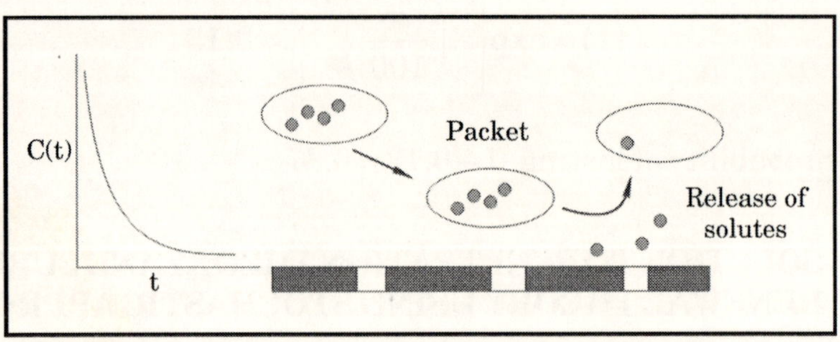

Fig. 8.3: Exponential distribution of packets over the membrane

Therefore, the distribution of packets with time of retention over the membrane surface can be written as,

$$\zeta(t) = \frac{\lambda^t \exp(-\lambda)}{t!} \tag{8.14}$$

Where, λ is the rate at which the packets have the probability to be retained on the surface of the membrane after advancing towards the membrane? With low rate, the probability of retention will be maximum, as the reverse diffusional gradient is minimum. While with high rate the packets or truly speaking the cluster retention will be minimum, as most of the microsolutes will pass through the membrane. As the time progresses the retention probability will be maximum because of the sticking of the microsolutes within the deposited macrosolutes on the surface. Further increase in the time, reverse concentration gradient resists the solutes to come over the membrane. Fig. 8.4 shows the nature of the Poisson distribution, which helps to clear our idea.

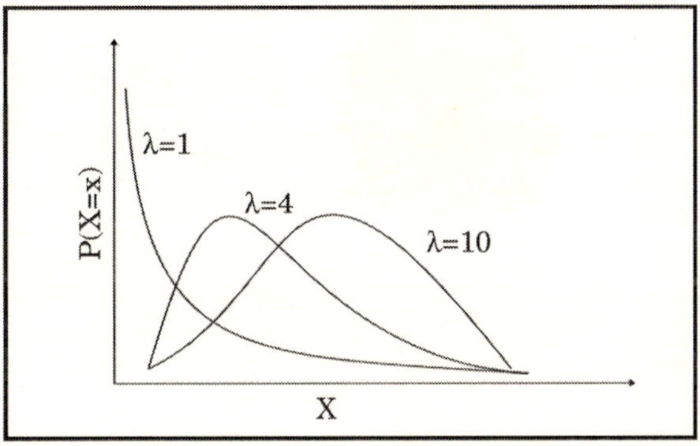

Fig. 8.4: Poisson distribution (probability P(X) of an event X)

Now from equation 8.12, one can say,

$$J_P(t) = J_C(t) - J_{BD}(t) \tag{8.15}$$

The convective mass flow for the microsolutes can be said as,

$$C(t)J_C(t) = k\left[C(z,t) - C_m(t)\right] \tag{8.16}$$

The back diffusional flux is primarily a function of time, which depends on the accumulation of the microsolutes over the membrane surface. Therefore, according to second law of Fick's diffusion, one can say,

$$C_m(t)J_{BD}(t) = -D\frac{\partial C}{\partial x} = -\int_0^{\infty}\left(\frac{\partial C(z,t)}{\partial t}\right)dz \qquad (8.17)$$

Hence, the permeate flux can be written as,

$$C_p(t)J_P(t) = k\left[C(z,t) - C_m(t)\right] + \int_0^{\infty}\left(\frac{\partial C(z,t)}{\partial t}\right)dz \qquad (8.18)$$

Now the second term on RHS of equation 8.18 can be expanded as,

$$C_p(t)J_P(t) = k\left[C(z,t) - C_m(t)\right] + \int_0^{\ell}\left(\frac{\partial C(z,t)}{\partial t}\right)dz + \int_{\ell}^{\infty}\left(\frac{\partial C(z,t)}{\partial t}\right)dz \qquad (8.19)$$

Where, ℓ is called the Prandtl mixing length. Hence, $\int_{\ell}^{\infty}\left(\frac{\partial C(z,t)}{\partial t}\right)dz = 0$.

Therefore, equation 8.19 becomes,

$$C_p(t)J_P(t) = k\left[C(z,t) - C_m(t)\right] + \int_{z=0}^{\ell}\left(\frac{\partial C(z,t)}{\partial t}\right)dz \qquad (8.20)$$

Now the term C(z, t) can be approximated as,

$$C(z,t) = C_{Bulk}\int_{z=0}^{\ell}\int_{t=0}^{\infty}\varphi(z,t)dzdt \qquad (8.21)$$

As, 'z' and 't' are two independent parameters, hence, the function j(z, t) can be written as the product of two functions given in equation 8.13 and 8.14.

$$\varphi(z,t)=\zeta(z)\psi(t) \tag{8.22}$$

$$C_m(t)=C(z,t)\Big|_{z=0}=C_{Bulk}\int_{t=0}^{\infty}\eta\frac{\lambda^t\exp(-\lambda)}{t!}dt \tag{8.23}$$

$$\left.\begin{aligned}\frac{\partial C(z,t)}{\partial t}&=C_{Bulk}\left(\eta\exp(-\eta z)\right)\frac{\partial\left(\dfrac{\lambda^t\exp(-\lambda)}{t!}\right)}{\partial t}\\&=C_{Bulk}\left(\eta\exp(-\eta z)\right)\left[\frac{\lambda}{t}\left\{\frac{(\lambda-1)^t\exp(-(\lambda-1))}{t!}-\frac{\lambda^t\exp(-\lambda)}{t!}\right\}\right]\\&=C_{Bulk}\left(\eta\exp(-\eta z)\right)\left[\frac{\lambda\exp(-\lambda)}{t(t!)}\left\{2.72(\lambda-1)^t-\lambda^t\right\}\right]\end{aligned}\right\} \tag{8.24}$$

Therefore, 8.20 becomes,

$$\begin{aligned}C_p(t)J_P(t)=kC_{Bulk}&\left[\int_{t=0}^{\infty}\int_{z=0}^{\ell}\left(\eta\exp(-\eta z)\right)\frac{\lambda^t\exp(-\lambda)}{t!}dzdt-\int_{t=0}^{\infty}\eta\frac{\lambda^t\exp(-\lambda)}{t!}dt\right]\\&+C_{Bulk}\left[\frac{\lambda\exp(-\lambda)}{t(t!)}\left\{2.72(\lambda-1)^t-\lambda^t\right\}\right]\int_{z=0}^{\ell}\left(\eta\exp(-\eta z)\right)dz\end{aligned} \tag{8.25}$$

Now, rearranging equation 8.25, we will have,

$$J_p(t)C_p(t) = kC_{Bulk}\left\{\begin{matrix}\left[\left[\int_{t=0}^{\infty}\int_{z=0}^{\ell}(\eta\exp(-\eta z))\frac{\lambda^t\exp(-\lambda)}{t!}dzdt - \int_{t=0}^{\infty}\eta\frac{\lambda^t\exp(-\lambda)}{t!}dt\right]\right] \\ +\left[\frac{\lambda\exp(-\lambda)}{t(t!)}\left\{2.72(\lambda-1)^t-\lambda^t\right\}\right]\int_{z=0}^{\ell}(\eta\exp(-\eta z))dz\end{matrix}\right\}$$

(8.26)

Now, mass balance against the membrane module can be written as follows,

$$V_0C_{Bulk} = \frac{d(VC(z,t))}{dt} + A_{Mem}J_p(t)C_p(t) \quad (8.27)$$

Combining equation 8.24, 8.26 and 8.27, we get,

$$1-\frac{kA_{Mem}}{V_0}\left\{\begin{matrix}\left[\left[\int_{t=0}^{\infty}\int_{z=0}^{\ell}(\eta\exp(-\eta z))\frac{\lambda^t\exp(-\lambda)}{t!}dzdt - \int_{t=0}^{\infty}\eta\frac{\lambda^t\exp(-\lambda)}{t!}dt\right]\right] \\ +\left[\frac{\lambda\exp(-\lambda)}{t(t!)}\left\{2.72(\lambda-1)^t-\lambda^t\right\}\right]\int_{z=0}^{\ell}(\eta\exp(-\eta z))dz\end{matrix}\right\}$$

$$-\tau(\eta\exp(-\eta z))\left[\frac{\lambda\exp(-\lambda)}{t(t!)}\left\{2.72(\lambda-1)^t-\lambda^t\right\}\right] = Z(\lambda,\eta)$$

(8.28)

In case with the surface renewal model, the renewal rate is not constant with time and can be approximated with the two parameter distribution – Weibull distribution.

Therefore equation 8.28 can be approximated as,

$$1-\frac{kA_{Mem}}{V_0}\left\{\begin{aligned}&\left[\left[\int_{t=0}^{\infty}\int_{z=0}^{\ell}\left(\eta\exp(-\eta z)\right)\frac{\lambda^{t}\exp(-\lambda)}{t!}dzdt-\int_{t=0}^{\infty}\eta\frac{\lambda^{t}\exp(-\lambda)}{t!}dt\right]\right.\\&\left.+\left[\frac{\lambda\exp(-\lambda)}{t(t!)}\left\{2.72(\lambda-1)^{t}-\lambda^{t}\right\}\right]_{}^{\ell}\int_{z=0}^{\ell}\left(\eta\exp(-\eta z)\right)dz\right]\end{aligned}\right\}$$
$$-\tau\left(\eta\exp(-\eta z)\right)\left[\frac{\lambda\exp(-\lambda)}{t(t!)}\left\{2.72(\lambda-1)^{t}-\lambda^{t}\right\}\right]_{}^{}=Z(\lambda,\eta)$$

(8.29)

8.2.1. Algorithm

Step 1: Guess *initial values* for λ and η.

Step 2: Note down the values for 'k', 'A_{Mem}', 'V_0' and 'V'.

Step 3: Calculate τ.

Step 4: *Define* tolerance.

Step 5: Integrate using Simpson's rule for the limit for 't' and 'z'.

Step 6: Evaluate the function Z (λ, η).

Step 7: If the evaluated function will be less than or equal to tolerance print values for λ and η. Else go to Step 5 with the *new guesses* for λ and η.

8.3. CHAPTER IN BRIEF

- Concept of statistical parameters.
- Concept of penetration theory and surface renewal theory.
- Application of statistical parameters in these theories.
- Mathematical analysis of membrane separation process with the aforesaid theories.

Chapter 9

FUZZY PREDICTIVE MODEL FOR MEMBRANE SEPARATION PROCESS

The critical reasoning behind the explanation to any real processes could be explicitly described through different uncertainties that are difficult to ascribe through conventional mathematical approaches. In case of membrane separation process, the uncertainties lie to the formation of cake layer over the membrane, which depends on either the interaction between the components to be separated with the membrane material or the geometry of the modules. As a case study given in chapter 2, where we are separating proteins from waste stream the membrane has a strong interaction with the solutes present in the stream. They are primarily composed of carbohydrates, proteins, salts etc. Now the carbohydrates, if the concentration is very high enough within the stream, carbohydrates may stick over the membrane surface even with low fouling membrane like polyether sulfone (PES) (Sen *et al.*, 2011b). Hence, approaching with the mass, momentum and energy balance to formulate the system is a bit difficult task here as much of the chemistry needs to be understood. While, the method described in the previous chapter might be considered as a viable scenario in order to formulate the process. However, the formulation is quiet cumbersome and requires a critical thinking on the mathematical analysis. In view of this, the most convenient scheme is the pattern mapping modelling like artificial neural network (ANN), where we need to rely much on the experimental information and their accuracy. With ANN such accuracy of the experimental outcomes might include a crucial scenario and give unpredictable values of the model outcomes for the membrane separation process. Now to avoid such unexpected occurrences of errors in model one can associate the qualitative statement with the quantitative one. This demand raises the question on fuzzy requirement in describing a membrane process mathematically. In this chapter, we will have a brief description on what we understand about pattern mapping design followed by an understanding of fuzzy system. Next we will try to evaluate these two approaches in modelling a membrane separation process and their shortcomings so far.

9.1. PATTERN MAPPING DESIGN

In pattern mapping design, we are matching cards of same colour and it is called mapping. Let us take one example of an apparels' shop, where the shopkeeper is arranging different grades of shirts in packets labelled with different price. Say, he had received grade A, B and C. The price of grade A is $10, grade B is $20 and grade C is $25 (Fig. 9.1). Accordingly, he has to store the apparels in the respective warehouses (as per arrow head). If he could actually do this, it is a tremendous trend that will be followed by each worker working under him and his training to the worker towards the selection of warehouse will be called perfect. If he made something wrong in selecting the correct warehouse, his worker will also imitate the same thing as they were trained. This is one of the shortfalls with the pattern mapping system. If the recognition of pattern will be done wrongly then the whole system fails and storage of apparels will be chaotic. With the membrane separation, once the experimental set-up will give accurate results, the mapping of the separation outcome with the experimental parameters will be good and hence, the formulation. Artificial neural network (ANN) is one of such pattern mapping example, where the system is visualised as neuron works.

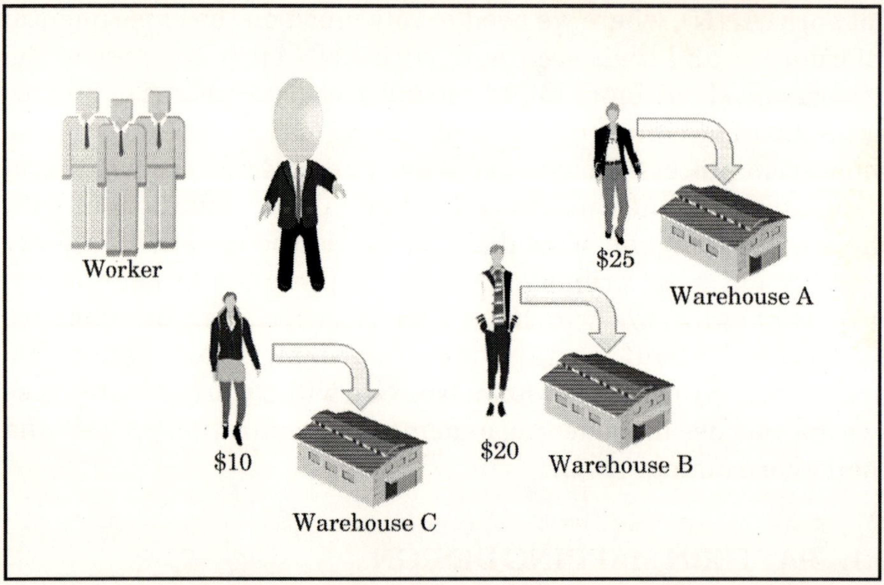

Fig. 9.1: Segregation of storage based on prices

9.1.1. History of Artificial Neural Network

The concept of artificial neural networks came in 1943[1], when Warren McCulloch, a neuro-physiologist, and Walter Pitts, a young mathematician, wrote a paper on how neurons might work. Adopting this concept of neurons and their function, a book was written by Donald Hebb in 1949 called *'The Organization of Behavior'*. In 1950s, Nathanial Rochester from the IBM research laboratories led the first effort to simulate a neural network. However, their first attempt failed followed by a successful subsequent attempt. In 1956, the Dartmouth Summer Research Project on Artificial Intelligence initiates the application of both artificial intelligence and neural networks. In the subsequent years after the Dartmouth Project, Frank Rosenblatt, a neuro-biologist of Cornell, began work on the Perceptron. The Perceptron, which was the outcome of this research, was built in hardware and it is the oldest neural network still in use today. The perceptron computes a weighted sum of the inputs, subtracts a threshold, and passes one of two possible values out as the result. In 1959, Bernard Widrow and Marcian Hoff of Stanford developed models, called ADAptive LINear Elements (ADALINE) and Multiple ADAptive LINear Elements (MADALINE). MADALINE was considered as the first neural network to be applied to a real world problem and is still in commercial use.

Unfortunately, a fear was set, with the success of the AI, on the effect of "thinking machines" on mankind. These fears, combined with unfulfilled, outrageous claims, ceased the research on artificial neural network and this period of stunted growth lasted through 1981. In 1982, John Hopfield of Caltech presented a paper to the national Academy of Sciences. Hopfield's approach was not to simply simulate brains but to create useful devices. He showed, with crisper clarity and mathematical analysis, how networks could work and what they could do. During the year 1985 to till 1989 several works had been going on the artificial neural network showing its importance in industrialisation and money flow. In 1990, US Department of Defence Small Business Innovation Research Program named 16 topics, which specifically targeted neural networks with an additional 13 mentioning the possible use of neural networks.

9.1.2. Topology of Neural Network

The topology of artificial neural network is quiet similar to our neural system where the action-reaction is mapped through brain (Fig. 9.2). Once the fire breaks out, the man relaxing there feels the heat and escapes. Now, the instruction to escape comes from brain, which one is transmitted through neurons and ultimately reaches there in brain. The man escapes because he was trained earlier that during fire one should run away in order to save life. This is how exactly the neural network works.

There is a receptor nodes' layer called input layer which receives the signal and transmits it to a layer called hidden layer through the connector. The signal is processed over here and

Fig. 9.2: Fire escape by the trained people

outcome will be delivered to output layer through connector again for response. Fig. 9.3 shows a typical structure for the neural network (Sen *et al.*, 2011a). Initially, in order to simulate a process, we have to provide a set of input conditions for which the output will be known. This is called '*training*' of the network and this is how we are teaching the network to response accurately. Once, the network will be trained properly the network can show the response for any set of conditions within the *limit*, until and unless we have a proper scenario within the hidden layer for which network can extrapolate. Now, the network is predicting response when the true values of the responses are known and we are finding out the deviation between these two responses in order to evaluate the training process. This is called '*validation*'. Finally, we will give some inputs for which we need to predict the responses without knowing the true values for those. This is called '*testing*'. The layers consist of nodes receiving the signal from the previous layer (as shown in Fig. 9.3). Nodes in the input layer are equal to the number of parameters for which we want to assess the response. Hidden layer(s) nodes depend on the complexity of the system and the nature of the neural network architecture we are using. Output layer nodes depend on the number of responses from the process we want to assess. Each node from one layer will be connected with the each of the subsequent layers' nodes and the connections are assigned with the weightages called '*weights*'. The hidden layer(s) nodes and output layer nodes are having biases, which will ensure that there should not be any '*zero*' value assigned to the nodes. Now, let us start the process with an example. A plant is there, which separates proteins from whey solution to manufacture whey protein concentrate using membrane separation process. Now, protein is something which is a macromolecules formed with amino acids as their unit. Based on the molecular weight of the proteins we have selected a specific MWCO for the membrane. The operating parameters that are adjusted to separate the proteins are pH and temperature, as we know both of these conditions have considerable effect on the structure of the proteins and thereby, their separation. We are trying to understand the average flux from the membrane after say 3 hours running the membrane setup. We have tested it on a bench scale before implementing it in large scale.

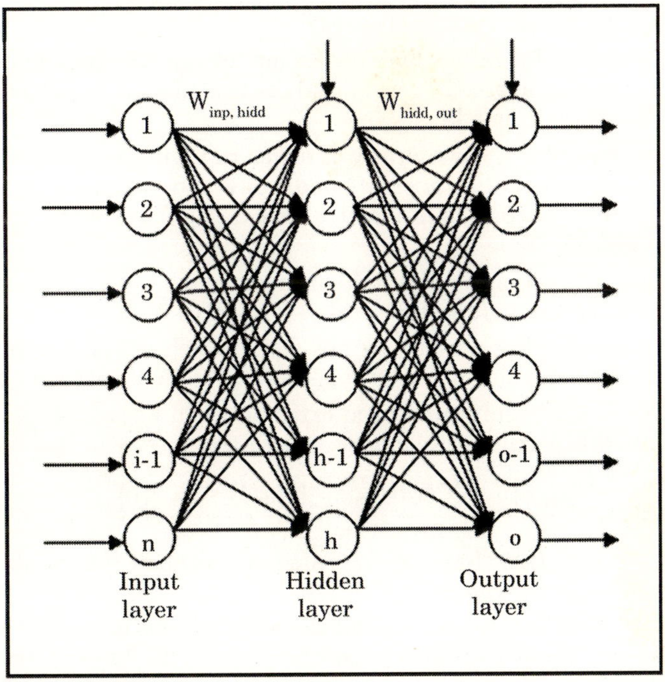

Fig. 9.3: Artificial neural network architecture (*Reprinted after the permission taken from Desalination, 273, 2011, 168-178*)

Condition 1: pH_i; i=1,2,3,……, N

Condition 2: T_i; i=1,2,3,……, N

Response 1: $J_{avg,I}$; i=1,2,3,……, N

Now out of this 'N' number of observations, we have taken 'n' numbers for the training of the network and 'N-n' numbers for the validation of the network. We are using simple feed-forward back-propagation network, where the flow of the prediction will be moving ahead. The error minimisation during simulation will be done from the output layer towards the input layer. The algorithm for the simulation is given in Tables 9.1 and 9.2.

Table 9.1: Training algorithm for feed-forward back-propagation network

Step	*Events for training*
1.	Note the observations in pH, T and J_{avg} array of size N.
2.	Isolate 'n' number of observations.
3.	Assign initial values for weights for the input to hidden (W_{ih}), biases on hidden layers (b_h), weights between hidden to output (W_{ho}) and biases on output layers (b_o).
4.	Assign i=2 *i.e.*, the number of nodes in input layer, which is exactly same with the number of conditions and o=1, *i.e.*, the number of nodes in output layer, which is exactly same with the response number.
5.	Assign function for the hidden layer which is say linear function of input to the hidden layer ($=b_h+\Sigma(W_{1h}pH_i+W_{2h}T_i)$).
6.	Assign function for the output layer which is say linear function of input to the output layer ($=b_o+\Sigma(W_{h1}f(b_h+\Sigma(W_{1h}pH_i+W_{2h}T_i)))$).
7.	Check the absolute average error between known observations and network predicted response: $\mathrm{MSE} = \frac{\left[J_{avg,i} - \left\{b_O + \sum W_{h1} f\left(b_h + \sum\left(W_{1h}pH_i + W_{2h}T_i\right)\right)\right\}\right]^2}{n}$
8.	If MSE is less than the tolerance, please call the validation model or else go to step 9.
9.	This MSE is a function of all the weights and biases of the network. Hence, apply appropriate mathematics (like Levenberg-Marquardt algorithm) to adjust the weights and biases of the network. Initially, the output to hidden layer connections is adjusted followed by hidden to input layer connections. Go to step 5 after adjustment.

Table 9.2: Validation algorithm for feed-forward back-propagation network

Step	Events for training
1.	Note the observations in pH, T and J_{avg} array of size N-n.
2.	Assign trained values for weights for the input to hidden (W_{ih}), biases on hidden layers (b_h), weights between hidden to output (W_{ho}) and biases on output layers (b_o).
3.	Assign i=2 *i.e.*, the number of nodes in input layer, which is exactly same with the number of conditions and o=1, *i.e.*, the number of nodes in output layer, which is exactly same with the response number.
4.	Assign function for the hidden layer which is say linear function of input to the hidden layer ($=b_h+\Sigma(W_{1h}pH_i+W_{2h}T_i)$).
5.	Assign function for the output layer which is say linear function of input to the output layer ($=b_o+\Sigma(W_{h1}f(b_h+\Sigma(W_{1h}pH_i+W_{2h}T_i)))$).
6.	Check the absolute average error between known observations and network predicted response: $\text{MSE} = \dfrac{\left[J_{avg,i} - \left\{b_O + \sum W_{h1} f\left(b_h + \sum\left(W_{1h} pH_i + W_{2h} T_i\right)\right)\right\}\right]^2}{n}$
7.	If MSE is less than the tolerance, the conclusion will be the training was done properly or else we need to train the network again with a new set of values.

However, it is clear from the analysis of the algorithm that the training is one of the important steps and if there will be any erroneous training of the dataset, the prediction will be wrong. Therefore, it is an absolute necessary to device a model which could predict the system quantitatively as well as qualitatively. If we just look at the artificial neural network, it is a type of binary system, where it tells that if the training information will be correct the network will predict correctly (*i.e.*, performance will be 1). On the contrary, if the training will be incorrect because of selection of wrong information the prediction will be incorrect (*i.e.*, performance will be 0). We do not make any qualitative statement in order to evaluate the process response. Let say, in the example if the pH will be higher one the quaternary structure of the protein will be unveiled and will be converted to tertiary, secondary and subsequently linear. Once, the structure will be linear, the molecular weight doesn't affect the separation and due to linear orientation even the higher molecular weight may pass through the membrane. Therefore, qualitative statement on the process performance will provide an additional benefit to assess the process response. This can be sorted out using fuzzy logic, where it tells the fractioned possibilities between 0 and 1. Combinatorial approach with both fuzzy and neural network will help to judge the process better. In the subsequent sections we will first describe fuzzy set and then its application with the neural system.

9.2. BASICS OF FUZZY SET

Professor Lotfi Zadeh (Zadeh, 1965) at the University of California, Berkley in 1965 proposed fuzzy set theory. In classical set theory the number either belongs to or not belongs to once we will go for the addition of two sets given by AB, called '*union*', where A and B are two sets of real numbers. Therefore, AÈB can be depicted as Fig. 9.5 and the mathematical expression for the AÈB is

$$N(A \cup B) = N(A)+N(B)-N(A \cap B) \tag{9.1}$$

$A \cap B$ depicts the concept that the number has to be present in both A and B, called '*intersection*' (see Fig. 9.4).

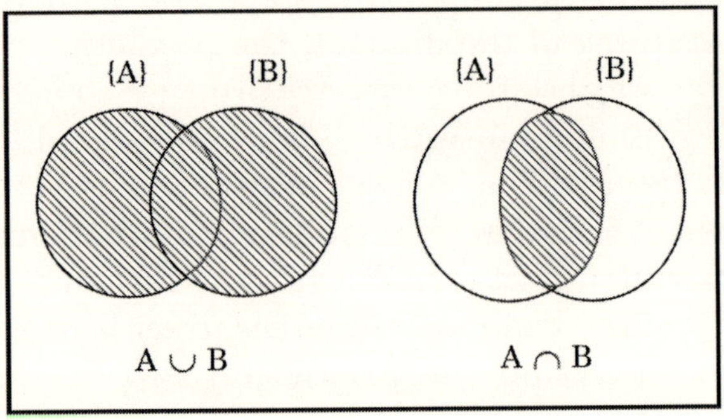

Fig. 9.4: Representation of union and intersection of set A and set B

Let us say in a class of 50 students 30 students are girls and 20 students are boys. Among the boys and girls 10 students got more than 90 marks in mathematics. How many boys and girls got less than 90 marks in mathematics? Now applying the formula above, we can say, this should be 30+20–10=40 students, when 'A' represents the girls' set and 'B' represents the boys' set. Now, in this case according to the problem the system tells two cases only, where the students either got more than 90 or less than 90.

Fuzzy set is capable of handling the situation, where it tells the demarcation percentage for every marks slab. Say 60% got more than 90, 20% got 80 to 90 ranges and thus so on. In fuzzy we have a membership function, which is denoted as $\mu_X(x) \in \langle 0,1 \rangle$. Now, what it actually says? It says the membership of the element 'x' within set 'X' will be in between 0 and 1 any value. Thus the membership function has the following properties (Yager, 1986):

(a) $\mu_{U\text{-}X}(x) = 1 - \mu_X(x),\ x \in U$

(b) $\mu_{X \cup Y}(x) = \max\{\mu_X(x), \mu_Y(x)\}$ or $\mu_X \vee \mu_Y,\ x \in U$

(c) $\mu_{X \cap Y}(x) = \min\{\mu_X(x), \mu_Y(x)\}$ or $\mu_X \wedge \mu_Y,\ x \in U$ (9.2)

'U' is universe. Let us elaborate the three properties by the above mentioned example. Qualitatively, we are distinguishing the students as the students got more than 90 are outstanding, in between 80 to 89 are excellent, in between 70 to 79 are very good, in between 60 to 69 are good, in between 50 to 59 are average, in between 40 to 49 are poor and less than 40 are failed. So we have seven memberships – '*outstanding*', '*excellent*', '*very good*', '*good*', '*average*', '*poor*' and '*failed*'. So now we have wide distribution for merits and this is good to assess the students' quality. We have the following distribution of students:

Marks	**>90**	**80–89**	**70–79**	**60–69**	**50–59**	**40–49**	**<40**
Boys (%)	10	20	30	20	10	5	5
Girls (%)	5	30	20	20	15	9	1

From the property the percentage of students got less than 50, will be max {0.05,0.05,0.09,0.01}=0.09 (property b). Percentage of students got more than 50, will be 1-0.09=0.91 (property a). Percentage of students got *at least* in between 50-59=min {0.1,0.15}=0.1 (property c) and percentage of students got in between 50-59=max {0.1,0.15}=0.15 (property b).

Now in this example the universal set (*universe of discourse*) is students for which the subset will be boys and girls. Let us elaborate the idea to a more extent, when we are saying about set (U) of students of school 'Xavier's' (x_1) and students of school 'Delhi Public School' (x_2). Say B is the subset of set U, which is saying about the male members among the students in both of the school. Now if both the schools are boys' schools then either the boys belong to the schools or not. This means the set possesses either '1' or '0' and this is called *crisp set*. Now if the schools are co-educational then say for Xavier's the boys are 30% of the student population and for the second one the boys are 60% of the student population. We may write the sets as below:

$$U=\{x_1,x_2\};\ B=\{0.3\,|\,x_1,0.6\,|\,x_2\}\text{ or }B=\{x_1\,|\,0.3,x_2\,|\,0.6\} \tag{9.3}$$

B(x) says the membership function of B and $B(x_1)$ is the membership grade of x_1 in the subset B.

If at least one membership grade of an element belongs to the universal set, is equal to 1 then the subset is called *normal* to the universal set. Else the set is *subnormal*. In the previous example B is subnormal fuzzy set. Now if Xavier's is boys' school then $B(x_1)=1$ and B will be normal fuzzy set.

Now maximum gradeship is called the *height* of fuzzy set and is written as height (B)=0.6. *Support* of B, written as Supp(B) is the crisp set of elements for which the gradeship will be nonzero. $Supp(B)=\{x_1,x_2\}$. Now, if Xavier's will be girls' school then B can be written as either $B=\{0\,|\,x_1,0.6\,|\,x_2\}$ or $B=\{x_1\,|\,0,x_2\,|\,0.6\}$. Here $Supp(B)=\{x_2\}$.

Now from the previous example we are again saying that the quality of the boys are assessed by another fuzzy set Q, which is given as 10% of the boys are good in academics from Xavier's and

5% of the boys are good in academics from Delhi Public School. Therefore, we can say 10% of $0.3x_1$ and 5% of $0.6x_2$ are good. Thus

$$Q=\{0.03\,|\,x_1, 0.03\,|\,x_2\} \tag{9.4}$$

As $B(x) \geq Q(x)$, then we can say Q is a subset of B and it is written as $Q \subset B$.

Theorem: If A and B are two fuzzy subsets of X and if $C=A \cup B$ and $D=A \cap B$ then,

a) $D \subset C$

b) $A \subset C$ and $B \subset C$

c) $D \subset A$ and $D \subset B$ (9.5)

If A is a fuzzy subset of X, then complement or negation of A, denoted $\bar{A}$ is defined as the fuzzy subset $\bar{A} = X - A$, hence for each $x \in A$, $A(x) = 1 - A(x)$. In the previous example $\overline{B} = \{0.7\,|\,x_1, 0.4\,|\,x_2\}$ says about the girls in both of the schools[2].

One of the important parameters in fuzzy logic is the fuzzy relationship. Suppose A and B are two fuzzy subsets for two sets X and Y, then the cross product of A x B is the relationship T on XxY such that T=AxB and $T(x,y)=\min\{A(x), B(y)\}$. Let it be explained by an example said before. From Xavier's the boys and girls have scored 'A' grade, 'B' grade and 'C' grade. Similarly, the students from Delhi Public School also have scored the same range of grades. $S=\{0.3\,|\,b_X, 0.7\,|\,g_X, 0.6\,|\,b_D, 0.4\,|\,g_D\}$ is the subset of students from Xavier's and Delhi Public School.

Now, $SQ=\{0.3\,|\,b_{XA}, 0.6\,|\,b_{XB}, 0.1\,|\,b_{XC}, 0.5\,|\,g_{XA}, 0.4\,|\,g_{XB}, 0.1\,|\,g_{XC}, 0.4\,|\,b_{DA}, 0.4\,|\,b_{DB}, 0.2\,|\,b_{DC}, 0.6\,|\,g_{DA}, 0.3\,|\,g_{DB}, 0.1\,|\,g_{DC}\}$ is the subset of students' quality from both the schools. Subscript 'D' stands for Delhi Public School and 'X' stands for Xavier's. Therefore, the relationship set for the sets S and SQ will be,

---

[2] Demorgan's law: $\overline{A \cup B} = \overline{A} \cup \overline{B}; \overline{A \cap B} = \overline{A} \cap \overline{B}$

If $A \subset B$ then $\overline{A} \supset \overline{B}$ and $A-B= \varphi$ (Null Set)

S x SQ={$0.3|(b_X,b_{XA})$, $0.3|(b_X,b_{XB})$, $0.1|(b_X,b_{XC})$, $0.3|(b_X,g_{XA})$, $0.3|(b_X,g_{XB})$, $0.1|(b_X,g_{XC})$, $0.3|(b_X,b_{DA})$, $0.3|(b_X,b_{DB})$, $0.2|(b_X,b_{DC})$, $0.3|(b_X,g_{DA})$, $0.3|(b_X,g_{DB})$, $0.1|(b_X,g_{DC})$, $0.3|(g_X,b_{XA})$, $0.6|(g_X,b_{XB})$, $0.1|(g_X,b_{XC})$, $0.5|(g_X,g_{XA})$, $0.4|(g_X,g_{XB})$, $0.1|(g_X,g_{XC})$, ………… $0.1|(g_D,g_{DC})$}

Moreover, in many situations people are more likely to describe and situation, where they are saying the words '*more or less*', '*nearly*', '*about*' etc. When these words or phrases are suffixed or prefixed by a number, these are called '*fuzzy numbers*'. Suppose, John is a student of Xavier's and his parents are being called in the school by the principal. He is interrogating John's parents. See Fig. 9.5. John's parents had used two key words '*around 50*' and '*more or less 5*', which are called fuzzy numbers.

Fig. 9.5: Conversation between John's parents

Assume X and Y are two sets. Let f be a mapping from X into Y, *i.e.*, $f:X \rightarrow Y$ such that for each $x \in X$, $f(x) = y \in U$. Assume A is a fuzzy subset of X, we can define f(A) as a fuzzy subset of Y such that f(A)

$$= \bigcup_{x \in X} \{A(x) \mid f(x)\}$$

Example: Assume X = {1, 2, 3} and Y = {a,b} and f is defined by f(1) = a, f(2) = a, f(3) = b. Let A = {1 | 1, 0.3 | 2, 0.7 | 3} then: f(A) = {1 | a}∪ {0.3 | a}∪ {0.7 | b} = {1 | a, 0.7 | b}. (Yager, 1986)

9.3. APPLICATION OF FUZZY AND NEURAL NETWORK IN MEMBRANE SEPARATION MODELING

The primary consideration to model a membrane separation process is to identify the parameters on which the performance of the process is depending. Now say consider a process, where we are using UF membrane to produce whey protein concentrate from whey solutions. Now if we want to judge the performance of the process, we must know the purity of the concentrate and what should be the separation factor for the proteins. Now, once we will go with the solution, the solution contains carbohydrate, bivalent ions like Ca^{2+} and proteins. As we all know, fouling is an obvious incident with the membrane process, therefore the fouling will be a prime factor controlling the performance. In this case, we may identify two scenarios of fouling – one reversible and the one is the irreversible. Now with the UF membrane the ions and the carbohydrates will go through. One possibility here might be that during the travel through the tortuous path of the membrane pores, the molecules may get stuck at the surface. Further, the tertiary structure of the protein molecules may get open up and the primary linear structure will move through the membrane. This will also result in the pore plugging. Apart from that we may have adsorption of protein on the membrane surface (it depends on the membrane material). Thus, so many of other such uncertain parameters may attribute to the fouling of the membrane. Therefore, we need some qualitative statement. If we want to describe it theoretically or if we want to model it theoretically, it becomes difficult to predict the realistic situation. With fuzzy interface neural network model we can elaborate or rather in true sense extend the prediction domain by using some qualitative statement along with the pattern mapping. If we are going with the only pattern mapping network as discussed in section 9.1, it is difficult to quantify the issues that are truly qualitative. For example suppose we want to analyze a process, where we have a feed of whey solution. Now we want to understand which module is good among hollow fibre and spiral wound. We are changing the buffer pH, temperature of the process and the VCF. All the later ones are quantitative. But what we will consider once we are thinking of the modules? We need qualitative statements. We can say, *'If the module is hollow fibre and if the pH is 7 and if the temperature is 70°C and if the VCF is 1.5, then the purity will be 56%'*. However, instead of saying definite values for the quantitative parameters we can also say, *'If the module is hollow fibre and if the pH is around 7 and if the temperature is around 70°C and if the VCF is around 1.5, then the purity will be 56%'*. The below diagram shows the above memberships and the result comes out of this memberships.

Based on the above system we are setting some rules,

R1: If the module is hollow fibre and if the pH is around 7 and if the temperature is around 70°C and if the VCF is around 1.5, then the purity will be 56%.

R2: If the module is spiral wound and if the pH is around 3 and if the temperature is around 70°C and if the VCF is around 1.5, then the purity will be 86%.

R3: If the module is spiral wound and if the pH is around 3 and if the temperature is around 30°C and if the VCF is around 2, then the purity will be 36%.

R4: If the module is dead-end and if the pH is around 7 and if the temperature is around 50°C and if the VCF is around 3, then the purity will be 20%.

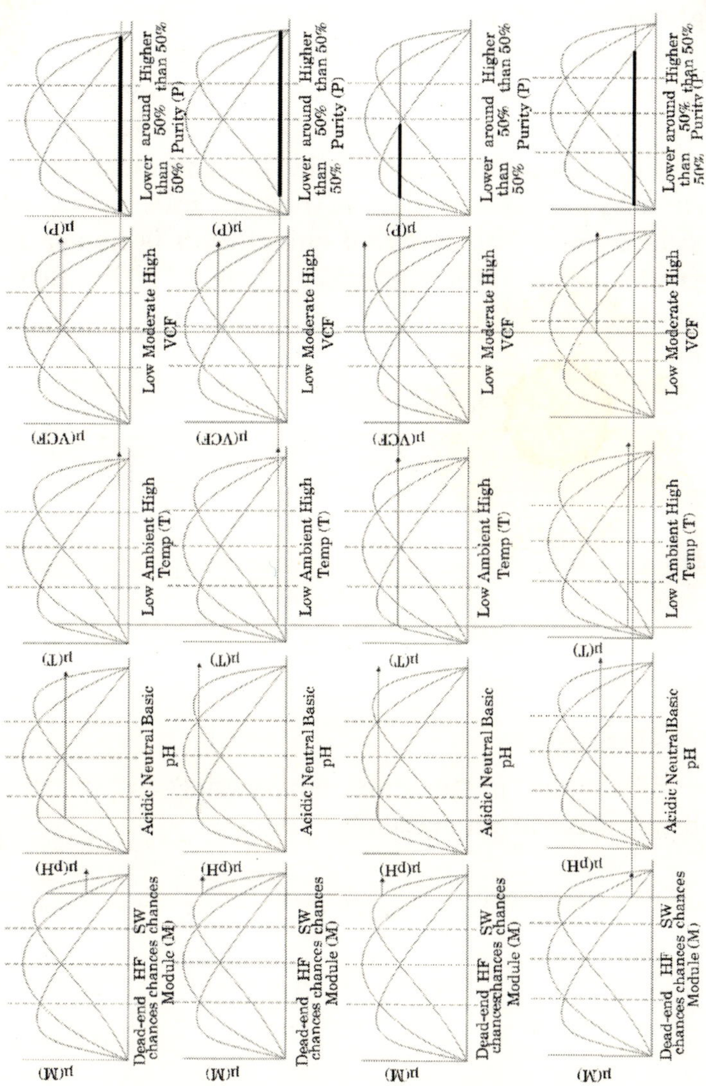

If the module is spiral wound and if the pH is around 2 and if the temperature is around 20°C and if the VCF is around 2, then what will be the purity

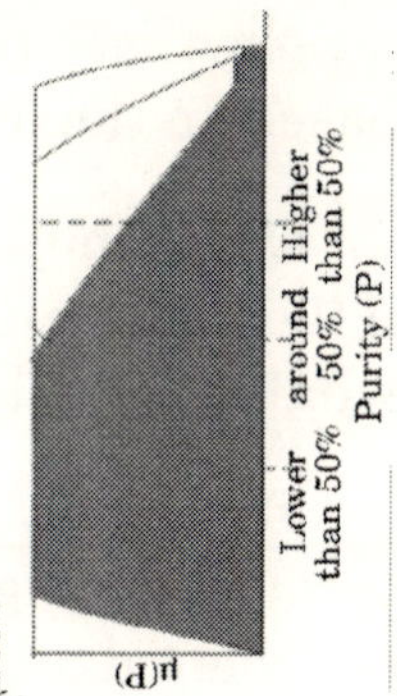

Fig. 9.6: Conclusion domain achieved after considering fuzzy rules

Now based on these rules if someone wants to predict the purity based on the following quality.

Q1: If the module is spiral wound and if the pH is around 2 and if the temperature is around 20°C and if the VCF is around 2, then what will be the purity.

The below Fig. 9.6 shows the outcome, which is evaluated through a weighted some of the above rules. The results shown in red colored area has a centroid, which gives the ultimate outcome of the process. During the marking of the rules on the membership function showing the conclusion, 'AND' (it is the intersection operator) gives the minimum value of the memberships for a particular rule among the rules, while 'OR' (it is the union operator) gives the maximum one. The centroid calculation is basically finding out the ratios between the statistical moment for the entire domain and the area under the curve.

$$x^* = \frac{\int \mu(x).x dx}{\int \mu(x) dx} \tag{9.6}$$

9.4. CHAPTER IN BRIEF

- Concept of pattern mapping model.
- Artificial neural network.
- Basics of fuzzy logic.
- Concept of fuzzy and pattern mapping model's application in membrane separation process.

Appendix

MATLAB CODES

A.1. CODE FOR JACOBI ITERATION

Problem: Find out the solution for the following set of simultaneous equations

$$10x_1 - x_2 + 2x_3 = 6$$
$$-x_1 + 11x_2 - x_3 + 3x_4 = 25$$
$$2x_1 - x_2 + 10x_3 - x_4 = -11$$
$$3x_2 - x_3 + 8x_4 = 15$$

```
%Solution to simultaneous equation

%using Jacobi iterative algorithm.

%Simultaneous equation

%-----------------------------------

%10x1-1x2+2x3=6

%-x1+11x2-x3+3x4=25

%2x1-x2+10x3-x4=-11
```

```
%3x2-x3+8x4=15
%Solve for x,y and z using Jacobi method.

function Jacobi
clc;

fprintf(1,'Enter dimension of the coefficient matrix:')
n=input('');
fprintf(1,'Enter maximum numer of iterations:')
maxIter=input('');
fprintf(1,'Enter tolerance:')
tolerance=input('');

%Solution matrix initialisation
for i=1:n
  fprintf(1,'Enter initial value x%d:',i);
  x_sol(i,1)=input('');
end

%Data entry for coefficient matrix
for k=1:3
  for i=1:n
```

```
        for j=1:n
          fprintf(1,'Enter value for a[%d][%d]:',i,j);
          a(i,j)=input('');
        end
        end
        if det(a)~=0
        break;
    elseif k==3
        fprintf(1,'Maximum number of entry reached');
    end
end
a
%Data entry for constant matrix
for i=1:n2
    fprintf(1,'Enter value for a[%d][%d]',i,n+1);
    a(i,n+1)=input('');
end

x_old=x_sol;
for iter=1:maxIter
    for i=1:n
        sum=0;
```

```
    for j=1:n
      if i<j
        sum=sum+x_old(j,1)*a(i,j);
      elseif i>j
        sum=sum+x_new(j,1)*a(i,j);
      end
    end
    x_new(i,1)=(a(i,n+1)-sum)/a(i,i);
  end
  abs(x_new-x_old)
  if abs(x_new-x_old)<= tolerance
    fprintf(1,'Solution=%g',x_new);
    break;
  end
end
fprintf(1,'Solution:x1=%g,x2=%g,x3=%g,x4=%g',x_new(1,1),x_
new(2,1),x_new(3,1));
end
```

A.2: CODE FOR NEWTON'S METHOD

Problem: Solve the following two simultaneous nonlinear equations using Newton's method.

$$f_1(x_1, x_2) = x_1^3 + x_2^2$$
$$f_2(x_1, x_2) = x_1^2 - x_2^3$$

```
function Newton_Method
%f1=x1^3+x2^2
%f2=x1^2-x2^3
clc;
for i=1:2
   fprintf(1,'Enter initial guess for x%d:',i);
   x(i,1)=input('');
end
iterMax=input('Enter maximum number of iterations:');
tol=input('Enter tolerance:');
for iter=1:iterMax
   funct_eval=func(x);
   Jacob_eval=Jac(x);
   xnew=x-inv(Jacob_eval)*funct_eval;
   if abs(xnew-x)<=tol
      break;
   end
   x=xnew;
end
fprintf(1,'Solution for x1:%.2g',xnew(1,1));
fprintf(1,'\nSolution for x2:%.2g',xnew(2,1));
end
```

```
function funct_eval=func(x)

funct_eval(1,1)=x(1,1)^3+x(2,1)^2;

funct_eval(2,1)=x(1,1)^2-x(2,1)^3;

end


function Jacob_eval=Jac(x)

Jacob_eval(1,1)=3*x(1,1)^2;

Jacob_eval(1,2)=2*x(2,1);

Jacob_eval(2,1)=2*x(1,1);

Jacob_eval(2,2)=-3*x(2,1)^2;

end
```

A.3. CODE FOR RUNGE KUTTA METHOD

Problem: Under appropriate conditions A decomposes as follows:

A -> R -> S

k_1=1/min k_2=0.01/min

R is to be produced from 10 litre/min of feed in which C_{A0} = 1 mol/ litre,

C_{R0}= Cs_0= 0.

The size of mixed flow reactor is 10 L. Find out the concentration profile with time assuming first order reaction using 4th order Runge-Kutta method. Assume the outlet flow rate is 9 litre/min.

```
function rungekutta
```

```
%
clc;
%
v=input('Enter CSTR volume in L:');
vi=input('Enter inlet flow rate in LPM:');
vo=input('Enter outlet flow rate in LPM:');
cao=input('Enter intial concentration for A (moles/L):');
cro=input('Enter intial concentration for R (moles/L):');
cso=input('Enter intial concentration for S (moles/L):');
k1=input('Enter 1st rate constant (/min):');
k2=input('Enter 2nd rate constant (/min):');
tf=input('Enter reaction time in mins:');
i=1;
ca=cao;
cr=cro;
cs=cso;
for t=1:0.1:tf
    dCadt=f1(v,vi,vo,cao,ca,k1);
    dCrdt=f2(v,vo,cr,ca,k2,k1);
    dCsdt=f3(v,vo,cs,cr,k2);
    can=rk1(v,vi,vo,cao,ca,k1)+ca;
    crn=rk2(v,vo,cr,ca,k2,k1)+cr;
```

```
  csn=rk3(v,vo,cs,cr,k2)+cs;
  mca(i)=can;
  mcr(i)=crn;
  mcs(i)=csn;
  i=i+1;
  ca=can;
  cr=crn;
  cs=csn;
end
t=1:0.1:tf
plot(t,mca,'k');
hold on
plot(t,mcr,'r');
hold on
plot(t,mcs,'g');
hold off
end
function [dCadt]=f1(v,vi,vo,cao,ca,k1)
   dCadt=(vi*cao-vo*ca)/v-k1*ca;
end
function [dCrdt]=f2(v,vo,cr,ca,k2,k1)
   dCrdt=-vo*cr/v-(k2*cr-k1*ca);
```

```
end
function [dCsdt]=f3(v,vo,cs,cr,k2)
   dCsdt=-vo*cs/v+k2*cr;
end
function [sum1]=rk1(v,vi,vo,cao,ca,k1)
   K0=0.1*f1(v,vi,vo,cao,ca,k1);
   K1=0.1*f1(v,vi,vo,cao,ca+K0/2,k1);
   K2=0.1*f1(v,vi,vo,cao,ca+K1/2,k1);
   K3=0.1*f1(v,vi,vo,cao,ca+K2/2,k1);
   sum1=(K0+2*K1+2*K2+K3)/6;
end
function [sum2]=rk2(v,vo,cr,ca,k2,k1)
   K0=0.1*f2(v,vo,cr,ca,k2,k1);
   K1=0.1*f2(v,vo,cr+K0/2,ca,k2,k1);
   K2=0.1*f2(v,vo,cr+K1/2,ca,k2,k1);
   K3=0.1*f2(v,vo,cr+K2/2,ca,k2,k1);
   sum2=(K0+2*K1+2*K2+K3)/6;
end
function [sum3]=rk3(v,vo,cs,cr,k2)
   K0=0.1*f3(v,vo,cs,cr,k2);
   K1=0.1*f3(v,vo,cs+K0/2,cr,k2);
   K2=0.1*f3(v,vo,cs+K1/2,cr,k2);
```

```
K3=0.1*f3(v,vo,cs+K2/2,cr,k2);

sum3=(K0+2*K1+2*K2+K3)/6;

end
```

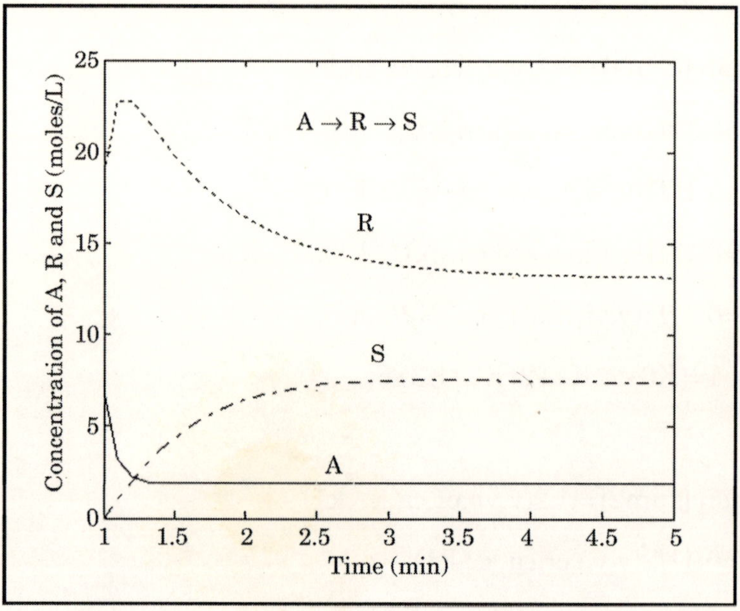

Fig. A.3: Concentration profile for A, R and S for a reaction time of 5 mins

A.4. CODE FOR TDMA

Problem statement: A property is varying along the space domain of length 5 unit, from its initial value of 3 units. The boundary is fixed with the initial boundary value 7 units to a final value of 5 units at any time 't'. The process is unsteady one and following the equation $\frac{\partial u}{\partial t} = 4\frac{\partial u^2}{\partial^2 x}$. Generate the property profile for a time limit of 3.

```
%PDE using TDMA
%Problem
%————
%du d2u
%— =4 ——
%dt dx2
%-------------------------------------------------------------------------------------
function TDMA
clc;
flag=0;
fprintf(1,'Enter the length of the space doamin:')
l=input('');
fprintf(1,'Enter the time scale:');
t=input('');
fprintf(1,'Enter the diffusivity term:');
k=input('');
for k=1:3
   fprintf(1,'Enter step size for length scale:');
   dell=input('');
   fprintf(1,'Enter step size for time scale:');
   delt=input('');
   lamda=k*delt/(dell^2);
```

```
    if lamda<=0.5
        break;
    elseif k==3
        flag = 1;
        display('Maximum number of inputs reached...');
    end
end
if flag==0
    nx=(l/dell)-2;
    nt=t/delt;
    fprintf(1,'Enter initial value for u(x,0):')
    u(1,1:nx+1)=input('');
    fprintf(1,'Enter value for u(0,t):')
    ulb=input('');
    fprintf(1,'Enter value for u(l,t):')
    uub=input('');
    %Coefficient matrix formation
    f=zeros(nx+1,nx+1);
    for xmr=1:nx+1
        for xmc=1:xmr+1
        if xmc < xmr-1
            continue;
```

```
        end
        if xmc > nx+1
            break;
        end
        if xmc == xmr
            f(xmr,xmc)=1+(2*lamda);
        else
            f(xmr,xmc)=-lamda;
        end
    end
end
%Initial constant matrix formation
fc(1:nx+1,1)=u';
fc(1,1)= fc(1,1)+(lamda*ulb);
fc(nx+1,1)= fc(nx+1,1)+(lamda*uub);
xplot=zeros(nt+1,(l/dell)+1);
xplot(1:nt+1,1)=ulb;
xplot(1:nt+1,(l/dell)+1)=uub;
x=1:(l/dell)+1;
for j=1:nt+1
    xplot(j,2:nx+2) = u;
    plot(x,xplot,'k');
```

```
    [ft,fct]=transform_matrix(f,fc,nx);
    [fc,u]=solution_matrix(fc,fct,ft,nx);
    fc(1,1)= fc(1,1)+(lamda*ulb);
    fc(nx+1,1)= fc(nx+1,1)+(lamda*uub);
    end
end
end

function [ft,fct]=transform_matrix(f,fc,nx)
ft=zeros(nx+1,nx+1);
fct=zeros(nx+1,1);
for i=1:nx+1
    for j=1:nx+1
        if i==j
            ft(i,j)=1;
        end
    end
end
ft(1,2)=f(1,2)/f(1,1);
fct(1,1)=fc(1,1)/f(1,1);
for i=2:nx
    ft(i,i+1)=f(i,i+1)/(f(i,i)-(ft(i-1,i)*f(i,i-1)));
```

```
    fct(i,1)=(fc(i,1)-(fct(i-1,1)*f(i,i-1)))/(f(i,i)-(ft(i-1,i)*f(i,i-1)));
end
fct(nx+1,1)=(fc(nx+1,1)-(fct(nx,1)*f(nx+1,nx)))/(f(nx+1,nx+1)-
(ft(nx,nx+1)*f(nx+1,nx)));
end


function [fc,u]=solution_matrix(fc,fct,ft,nx)
ux(nx+1,1)=fct(nx+1,1);
for i=nx:-1:1
    ux(i,1)=fct(i,1)-(fc(i+1,1)*ft(i,i+1));
end
```

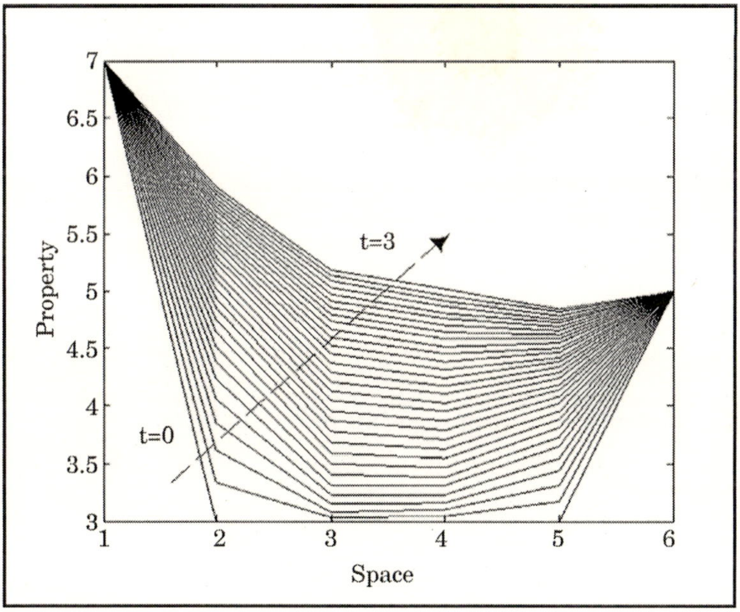

Fig. A.4: Showing the variation of property along the space domain at different time scale

```
fc=ux;
u=ux';
end
```

A.5. CODE FOR ADAPTIVE SIMPSON'S RULE

Problem: Solve $\int_{1}^{1.5} x^2 \ln x dx$ using adaptive Simpson's rule.

```
function adaptSimpson


fprintf(1,'Enter lower limit of the domain (x0):');
xal=input('');
fprintf(1,'Enter upper limit of the domain (xl):');
xbl=input('');
fprintf(1,'Enter stepsize (h):');
h=input('');
fprintf(1,'Enter tolerance:');
tol=input('');
fprintf(1,'Enter maximum number of levels:');
levelMax=input('');
%Initialisation
segment=2;
node=segment+1;
```

```
%Iteration start
for level=1:levelMax
    x(level,1)=xal;
    x(level,node)=xbl;
    for n=2:node-1
        x(level,n)=0.5*(x(level,n-1)+x(level,n+1));
    end
    j=1;
    for n=1:2:node-2
        xa=x(level,n);
        xb=x(level,n+2);
        xc=0.5*(xa+xb);
        Iab=(h/6)*(f(xa)+(4*f(xc))+f(xb));
        xd=0.5*(xa+xc);
        Iac=(h/6)*(f(xa)+(4*f(xd))+f(xc));
        xe=0.5*(xc+xb);
        Icb=(h/6)*(f(xc)+(4*f(xe))+f(xb));
        err(j)=abs(Iab-Iac-Icb);
        I(j)=(Iac+Icb)+(((Iac+Icb)-Iab)/15);
        j=j+1;
    end
    sume=0;
    sumI=0;
```

```
    l=length(err);
    for j=1:l
        sume=sume+err(j);
        sumI=sumI+I(j);
    end
    if sume<=15*tol
        sumI
        level
        break;
    end
    segment=segment*2;
    node=segment+1;
    count=1;
    for n=3:2:node-2
        x(level+1,n)=x(level,n-count);
        count=count+1;
    end
end
end

function result=f(x)
result=(x^2)*log(x);
end
```

Bibliography

Absolom, D.R., Zingg, W. and Neumann, A.W. (1987). Protein adsorption to polymer particles: Role of surface properties. *J. Biomed. Mater. Res.*, 21: 161–171. doi:10.1002/jbm.820210202

Abulencia, J.P. and Theodore, L. (2015). Open-Ended Problems: A Future Chemical Engineering Education Approach. Scrivener Publishing LLC., USA & John Wiley & Sons, Inc. Hoboken, USA.

Alexander, C. and Rietschel, E.T. (2001). Invited review: Bacterial lipopolysaccharides and innate immunity. *J. Endotoxin Res.,* 7: 167–202. doi:10.1177/09680519010070030101

Asatekin, A. and Mayes, A.M. (2009). Oil Industry wastewater treatment with fouling resistant membranes containing amphiphilic comb copolymers. *Environ. Sci. Technol.*, 43: 4487–4492. doi:10.1021/es803677k

Attard, P. (2009). The second entropy: A general theory for non-equilibrium thermodynamics and statistical mechanics. *Annu. Reports Sect.* "C" *Physical Chem.* 105: 63–173. doi:10.1039/B802697C

Blagg, C.R. (2015). The early history of dialysis for chronic renal failure in the United States: A view from seattle. *Am. J. Kidney Dis.*, 49: 482–496. doi:10.1053/j.ajkd.2007.01.017

Bouré, T. and Vanholder, R. (2004). Which dialyser membrane to choose? Nephrol. *Dial. Transplant.*, 19: 293–296. doi:10.1093/ndt/gfg508

Chang, I.-S., Field, R. and Cui, Z. (2009). Limitations of resistance-in-series model for fouling analysis in membrane bioreactors: A cautionary note. *Desalin. Water Treat.*, 8: 31–36. doi:10.5004/dwt.2009.687

Cimino, J.E., Brescia, M.J. and Aboody, R. (1962). Simple venipuncture for hemodialysis. *N. Engl. J. Med.*, 267: 608–609. doi:10.1056/NEJM196209202671207

Dickel, G. (1984). The variational principles of onsager and prigogine in membrane transport. Faraday Discuss. *Chem. Soc.*, 77: 157–168. doi:10.1039/DC9847700157

Dinarello, C.A. (1992). Cytokines: Agents provocateurs in hemodialysis? *Kidney Int.*, 41: 683–694.

Eknoyan, G., Beck, G.J., Cheung, A.K., Daugirdas, J.T., Greene, T., Kusek, J.W., Allon, M., Bailey, J., Delmez, J.A., Depner, T.A., Dwyer, J.T., Levey, A., Levin, N.W., Milford, E., Ornt, D.B., Rocco, M. V, Schulman, G., Schwab, S.J., Teehan, B.P. and Toto, R. (2002). Effect of dialysis dose and membrane flux in maintenance hemodialysis. *N. Engl. J. Med.*, 347: 2010–2019. doi:10.1056/NEJMoa021583

El-Wakil, H.S., Abou-Zeid, A.A., El-Gohary, I.E., Abou El-Seoud, N.A., Abou El-Yazeed, M.A. and El-Leithy, M.A. (2011). Effect of high flux versus low flux hemodialysis on serum beta-2 microglobulin, advanced oxidation protein products and protein carbonyl. *Alexandria J. Med.*, 47: 37–42. doi:http://dx.doi.org/10.1016/j.ajme.2011.04.004

Energy-Efficient Process for Using Membrane Technology to Treat and Recycle Agricultural Drainage Water [WWW Document], 2016. URL *http: / / www.energy.ca.gov / pier / portfolio / Content / Completed97to06 / Completedprior05plusEISG05 / Energy Efficient Process for.htm* (accessed 1.1.16).

Field, R.W., Wu, D., Howell, J.A. and Gupta, B.B. (1995). Critical flux concept for microfiltration fouling. *J. Memb. Sci.*, 100: 259–272. doi:http://dx.doi.org/10.1016/0376-7388(94)00265-Z

Furtos, G., Rivero, G., Rapuntean, S. and Abraham, G.A. (2016). Amoxicillin-loaded electrospun nanocomposite membranes for dental applications. *J. Biomed. Mater. Res. Part B Appl. Biomater.*, doi:10.1002/jbm.b.33629

Grossman, E. (2016). We've Changed a Life-Giving Nutrient into a Deadly Pollutant. How Can We Change it Back? Ensia.

Haldar, D., Sen, D. and Gayen, K. (2016). A Review on the Production of Fermentable Sugars from Lignocellulosic Biomass through Conventional and Enzymatic Route – A Comparison. *Int. J. Green Energy,* doi:10.1080/15435075.2016.1181075. doi:10.1080/15435075.2016.1181075

Hassan, B.I., Ennouri, M., Lafforgue, C., Schmitz, P. and Ayadi, A. (2013). Experimental Study of Membrane Fouling during Crossflow Microfiltration of Yeast and Bacteria Suspensions: Towards an Analysis at the Microscopic Level. Membranes (Basel). doi:10.3390/membranes3020044

Higbie, R. (1935). The Rate of absorption of a pure gas into a still liquid during short periods of exposure. *Trans. AIChE* 31, 365–389.

Hwang, S.-T. (2004). Nonequilibrium thermodynamics of membrane transport. *AIChE J.*, 50: 862–870. doi:10.1002/aic.10082

Isikgor, F.H. and Becer, C.R. (2015). Lignocellulosic biomass: a sustainable platform for the production of bio-based chemicals and polymers. *Polym. Chem.*, 6: 4497–4559.

Jenkins, D. and Wanner, J. (2014). Activated Sludge - 100 Years and Counting. IWA Publishing.

Kolff, W.J. (1965). First Clinical Experience with the Artificial Kidney. *Ann. Intern. Med.*, 62: 608–619. doi:10.7326/0003-4819-62-3-608

Laude-Sharp, M., Caroff, M., Simard, L., Pusineri, C., Kazatchkine, M.D. and Haeffner-Cavaillon, N. (1990). Induction of IL-1 during hemodialysis: Transmembrane passage of intact endotoxins (LPS). *Kidney Int.*, 38: 1089–1094.

Mallada, R. and Menéndez, M. (2008). Inorganic Membranes: Synthesis, Characterization and Applications, Membrane Science and Technology. Elsevier Science.

Martinez, M. and Vuylsteke, A. (2012). Extracorporeal membrane oxygenation in adults. *BJA Educ.*, 12: 57–61.

Metzger, L.E., Lee, J.K., Finer-Moore, J.S., Raetz, C.R.H. and Stroud, R.M. (2012). LpxI structures reveal how a lipid A precursor is synthesized. *Nat. Struct. Mol. Biol.*, 19: 1132–1138.

Mumtaz, A., Anees, M., Bilal, M. and Ibrahim, M. (2010). Beta-2 microglobulin levels in hemodialysis patients. *Saudi J. Kidney Dis. Transplant.*, 21: 701–706.

Nguyen, T., Roddick, F.A. and Fan, L. (2012). Biofouling of water treatment membranes: A review of the underlying causes, monitoring techniques and control measures. *Membranes* (Basel)., 2: 804–840.

Paul, W. (2006). Membranes for Haemodialysis, *In*: Somasundaram, P. and Hubbard, T. (*eds.*), Encyclopedia of Surface and Colloid Science, Volume 5. Taylor and Francis, USA, pp. 3565–3577.

Peek, G.J., Clemens, F., Elbourne, D., Firmin, R., Hardy, P., Hibbert, C., Killer, H., Mugford, M., Thalanany, M., Tiruvoipati, R., Truesdale, A. and Wilson, A. (2006). CESAR: conventional ventilatory support *vs* extracorporeal membrane oxygenation for severe adult respiratory failure. *BMC Health Serv. Res.*, 6: 163. doi:10.1186/1472-6963-6-163

Pervaporation developments (1992). Membr. Technol. pp. 6–8. doi:http://dx.doi.org/10.1016/0958-2118(92)90031-3

Plaza, A., Merlet, G., Hasanoglu, A., Isaacs, M., Sanchez, J. and Romero, J. (2013). Separation of butanol from ABE mixtures by sweep gas pervaporation using a supported gelled ionic liquid membrane: Analysis of transport phenomena and selectivity. *J. Memb. Sci.*, 444: 201–212. doi:http://dx.doi.org/10.1016/j.memsci.2013.04.034

Putatunda, S., Sen, D. and Bhattacharjee, C. (2015). Emulsified oily-waste water: Pretreatment with inorganic salts and treatment using two indigenous membrane modules. *RSC Adv.*, 5: 52676–52686. doi:10.1039/C5RA09621A

Putatunda, S., Sen, D., Sarkar, A., Chowdhury, R., Sarkar, D. and Bhattacharjee, C. (2014). Two indigenous high sheared membrane modules' performance expatiation for the ultrafiltration of polyethylene glycol. *RSC Adv.*, 4: 6435–6446. doi:10.1039/c3ra44117b

Qazi, S., Liepold, L.O., Abedin, M.J., Johnson, B., Prevelige, P., Frank, J.A. and Douglas, T. (2013). P22 viral capsids as nanocomposite high-relaxivity MRI contrast agents. *Mol. Pharm.*, 10: 11–17. doi:10.1021/mp300208g

Scribner, B.H. (2002). Medical dilemmas: The old is new. *Nat. Med.*, 8: 1066–1067. doi:10.1038/nm772

Sen, D., Roy, A., Bhattacharya, A., Banerjee, D. and Bhattacharjee, C. (2011a). Development of a knowledge based hybrid neural network (KBHNN) for studying the effect of diafiltration during ultrafiltration of whey. *Desalination*, 273: 168–178. doi:10.1016/j.desal.2010.10.038

Sen, D., Sarkar, A., Gosling, A., Gras, S.L., Stevens, G.W., Kentish, S.E., Bhattacharya, P.K., Barber, A.R. and Bhattacharjee, C. (2011b). Feasibility study of enzyme immobilization on polymeric membrane: A case study with enzymatically galacto-oligosaccharides production from lactose. *J. Memb. Sci.*, 378: 471–478. doi:10.1016/j.memsci.2011.05.032

Shuval, H. (2012). Water Renovation and Reuse. Academic Press Inc., UK

Szymczak, P. and Ladd, A.J.C. (2004). Stochastic boundary conditions to the convection-diffusion equation including chemical reactions at solid surfaces. *Phys. Rev. E*, 69: 36704.

Thomas, M., Moriyama, K. and Ledebo, I. (2011). AN69: Evolution of the World's First High Permeability Membrane, *In*: Saito, A., Kawanishi, H., Yamashita, A.C. and Mineshima, M. (*eds.*), Contributions to Nephrology: High-Performance Membrane Dialyzers (Vol. 173). Karger Medical and Scientific Publishers, Bassel (Switzerland), pp. 119–129.

Tijms, H.C. (2003). A First Course in Stochastic Models. John Wiley & Sons Ltd, UK.

Tiraferri, A. (2015). Membrane-based water treatment to increase water supply.

Vercauteren, S., Keizer, K., Vansant, E.F., Luyten, J. and Leysen, R. (1998). Porous ceramic membranes: Preparation, transport properties and applications. *J. Porous Mater.*, 5: 241–258. doi:10.1023/A:1009634305315

Von Sperling, M. (2007). Activated Sludge and Aerobic Biofilm Reactors, Biological Wastewater Treatment Series. IWA Publishing.

Wijmans, J.G. (Hans) and Baker, R.W. (2006). The solution–diffusion model: A unified approach to membrane permeation, *In*: Materials Science of Membranes for Gas and Vapor Separation. John Wiley & Sons, Ltd, pp. 159–189. doi:10.1002/047002903X.ch5

Yager, R.R. (1986). An introduction to Fuzzy set theory, *In*: Karwowski, W. and Mittal, A. (*eds.*), Applications of Fuzzy Set Theory in Human Factors. Elsevier B.V., Amsterdam, Netherland, pp. 29–39.

Zadeh, L.A. (1965). Fuzzy sets. *Inf. Control*, 8: 338–353.

Zhang, B., Shi, W., Yu, S., Zhu, Y., Zhang, R. and Li, L. (2015). Optimization of cleaning conditions on a polytetrafluoroethylene (PTFE) microfiltration membrane used in treatment of oil-field wastewater. *RSC Adv.*, 5: 104960–104971. doi:10.1039/C5RA23980J

Zhu, B., Kim, H.J., Na, Y.-H., Moon, I.-S., Connor, G., Maeda, S., Morris, G., Gray, S. and Duke, M. (2013). Temperature and Pressure Effects of Desalination Using a MFI-Type Zeolite Membrane. Membr. doi:10.3390/membranes3030155

Subject Index

A

B

C

I

J

L

M

N

O

P

R

S

T